Microbe-Assis
Phytoremediatio
Environmental Poll

Microbe-Assisted Phytoremediation of Environmental Pollutants

Recent Advances and Challenges

Vineet Kumar
Gaurav Saxena

CRC Press
Taylor & Francis Group
Boca Raton London New York

CRC Press is an imprint of the
Taylor & Francis Group, an **informa** business

AN A K PETERS BOOK

First edition published 2020
by CRC Press
6000 Broken Sound Parkway NW, Suite 300, Boca Raton, FL 33487-2742

and by CRC Press
2 Park Square, Milton Park, Abingdon, Oxon, OX14 4RN

© 2021 Taylor & Francis Group, LLC

CRC Press is an imprint of Taylor & Francis Group, LLC

ISBN: 978-0-367-33057-6 (hbk)
ISBN: 978-0-429-31780-4 (ebk)

Typeset in Times
by Nova Techset Private Limited, Bengaluru & Chennai, India

Visit the Taylor & Francis Web site at
http://www.taylorandfrancis.com

and the CRC Press Web site at
http://www.crcpress.com

This book is truly dedicated to our parents for their unfailing patience, contagious love, forgiveness, selflessness, endless support, and nurturing and educating us to date. Without them, we wouldn't be the people we are today.

Contents

Preface

The environment comprises air, water, and soil, which support the growth of all living organisms on the earth and provide basic necessities to carry out all the biological activities. However, worldwide contamination of water, soil, and air with hazardous organic and/or inorganic pollutants arising from anthropogenic sources as well as natural disasters poses several environmental and human health problems. The negative effects of pollutants on the environment and human health are diverse and depend on the nature and type of pollution. Therefore, there is an immediate global demand for remediation of environmental pollutants from soil, water, and air to counteract the adverse effects on human health and conserve the environment for our future generations. A wide range of methods based not only on the physical and chemical but also biological means have been available for the remediation of contaminated sites for decades, but these methods are energy-intensive and financially expensive, cause secondary air/soil/groundwater pollution, and can kill the pollutant-degrading microbial communities. Therefore, the conservation of our natural environment requires the development of sustainable environmental remediation technologies that promise through an economical and environmentally friendly way to restore the contaminated sites for human rehabilitation and agricultural production in an effective and more sustainable manner.

To safeguard the environmental and public health, there is an urgent need to develop some novel environmentally friendly remediation methods that help to overcome the adverse impacts of environmental pollution, and microbe-assisted phytoremediation has emerged as a novel holistic approach. It is a low-cost and eco-sustainable remediation approach that uses a broad range of plants and their associated microbes for the degradation and detoxification of environmental pollutants in contaminated environmental matrices (soil and water). Microorganisms and plants can metabolize (degrade/detoxify) and/or biotransform several recalcitrant pollutants either to obtain carbon and/or energy for growth and development or as co-substrates, thus converting them to simpler end products such as carbon dioxide, water, and chloride. Although the diverse metabolic capabilities of plants and associated microbes, and their interactions with hazardous organic and inorganic pollutants, have been revealed in the past few years, the detailed knowledge regarding the concepts and mechanisms of microbe-assisted phytoremediation of organic and inorganic pollutants in the different environmental matrices (soil and water) has been scattered and not available as a whole to researchers.

Therefore, the present book, entitled *Microbe-Assisted Phytoremediation of Environmental Pollutants: Recent Advances and Challenges*, has compiled the updated knowledge of microbe-assisted phytoremediation of potentially toxic environmental pollutants, such as toxic metals, pesticides, azo dyes, petroleum hydrocarbons, phenols, chlorophenols, etc. in a more comprehensive manner. This whole book is spread over 12 chapters and provides an overview of the latest research and development in different aspects of microbe-assisted phytoremediation for the ecological restoration and safety of public health. All the chapters in the book are written in a more comprehensive way and are meticulously prepared with fabulous figures and tables to make the information easier to understand, and are supported by an extensive list of references and URLs for readers interested in learning further details about the subject matter. Thus, in our opinion, this book is extremely useful to readers, and explicitly targeted as good teaching material for graduates as well as a reference for remediation practitioners and researchers, and will go some way to make this planet cleaner and greener. This book will be of great value to researchers, environmentalists, ecotoxicologists, environmental scientists and engineers, environmental science managers and administrators, policymakers and students at the master's and doctoral level, who wish to work on the microbe-assisted phytoremediation of pollutants for environmental sustainability.

Vineet Kumar
Gaurav Saxena

Acknowledgments

Microbe-Assisted Phytoremediation of Environmental Pollutants: Recent Advances and Challenges is the outcome of our dedicated efforts of almost two years that we took to complete this valuable book. During this journey, we recognized many individuals who directly or indirectly supported us during the writing of this precious book and bringing it into this unique world, many of whom deserve special mention. First and foremost, we would like to acknowledge and thank The Almighty God, Lord Shiva, for blessing us with enough energy, potential, strength, wisdom, and support, without whom nothing would be possible but with whom everything is!

On a very personal note, we are indebted to our family members for their patience, love, affection, and perseverance, as well as their constant encouragement and belief during the writing of this book. They wholeheartedly supported us in our overburdened schedule and stood with us during this long journey. Any success that we have achieved or will achieve in the future would not be possible without the love and kind support of our beloved families.

We are sincerely thankful to Prof. Ram Chandra and Dr. Ram Naresh Bharagava, Department of Environmental Microbiology, School of Environmental Sciences, Babasaheb Bhimrao Ambedkar University (A Central University), Lucknow, Uttar Pradesh, India, for their tremendous academic support and several wonderful opportunities that they provided during the course of our studies and, most importantly, supporting us in this unique field of environmental microbiology (bioremediation and microbe-assisted phytoremediation).

Our lifelong appreciation and gratitude also go to some important people at the Department of Environmental Microbiology of our prestigious institution, Babasaheb Bhimrao Ambedkar University (A Central University), Lucknow, Uttar Pradesh, India. We are wholeheartedly thankful to our colleagues and research fellows, namely Mr. Kshitij Singh, Mr. Adarsh Kumar, Mr. Ajay Kumar Singh, Mr. Jai Prakash, and Mr. Roop Kishor, for their constant support, encouragement, incredible ideas, and warm hospitality. The good times spent with them can never be forgotten and will be cherished throughout our lives.

We are most sincerely thankful to Professor Indu Sekhar Thakur at the School of Environmental Sciences, Jawaharlal Nehru University, New Delhi (India), for his advice, dedicated help, and moral support, and for providing us a fantastic facility in his laboratory (Environmental Microbiology and Biotechnology Laboratory) to complete this task. We are also thankful to all the faculty and staff members of our school for their help and support during our course of study. Special thanks are due to the funding agency, Department of Biotechnology (DBT), Government of India (GOI), New Delhi, for providing funding to carry out research work at the School of Environmental Sciences, Jawaharlal Nehru University, New Delhi (India).

We are extremely grateful to our publishing editors, Renu Upadhyay, senior acquisition/commissioning editor (Life Science) and Shikha Garg, editorial assistant (Life Sciences) at CRC Press, Taylor & Francis Group, for the execution of the publishing agreement, encouragement, support, and valuable advice/suggestions during the book project. We are thankful to Paul Boyd, our production editor at CRC Press, Taylor & Francis Group, Florida, and Kelsey Barham, our project manager at Nova Techset Private Limited, India, for the skillful organization and management of the entire book project in many ways (illustrations, graphics/book cover designing, copyediting, typesetting, proofing, etc.). Many thanks are due to Jyotsna Jangra, editorial assistant (Life Sciences), CRC Press, Taylor & Francis Group, India, for the kind invitation to write this wonderful book on behalf of the world-famous CRC Press (USA). Further, we are also thankful to the entire team at CRC Press, Taylor & Francis Group, Florida and India, who directly or indirectly contributed to the development and bringing of this valuable book into the world.

Further, we hope that this book will be of great value to researchers in the area of microbe-assisted phytoremediation and will go some way to make our planet safe and greener. In the end, we seek to learn more on the subject through the valuable comments, reviews, and suggestions from our readers, which can be directly sent to our e-mails: vineet.way18@gmail.com or drvineet.micro@gmail.com (Vineet Kumar) and gaurav10saxena@gmail.com (Gaurav Saxena). We shall be very happy to reply.

Authors

 **Vineet Kumar** currently works at the School of Environmental Sciences, Jawaharlal Nehru University, Delhi (India). He received his BSc (2008) in Biology and Chemistry from Meerut College, Meerut, and his MSc (2010) in Microbiology from Ch. Charan Singh University, Uttar Pradesh (UP), India. At the same university, he also earned MPhil (2011) in Microbiology. He joined the Department of Environmental Microbiology in 2012 at Babasaheb Bhimrao Ambedkar (A Central University) (BBAU), UP, India, where he completed his doctoral work (2018) in the field of Environmental Microbiology (specializing in bioremediation and metagenomics of distillery waste). Due to his outstanding academic contribution during his PhD, he received a merit certificate from the vice-chancellor of BBAU. His research focused on understanding the role and mechanism of rhizospheric bacterial communities in assisting metal uptake by wetland plants and the assessment of bioaugmentation and biostimulation approaches for bioremediation of hazardous industrial waste. He has published eight research articles in high-impact peer-reviewed journals published by Springer Nature, Elsevier, and Frontiers Media. He has also authored/co-authored four research articles published in conference proceedings, 28 book chapters, and six magazine articles on different aspects of bioremediation and phytoremediation of industrial waste pollutants. He is the editor of four books—*Phytoremediation of Environmental Pollutants* (CRC Press, USA); *Bioremediation for Environmental Sustainability: (Vol-I) Toxicity, Mechanisms of Contaminants Degradation, Detoxification and Challenges* (Elsevier, USA); *Bioremediation for Environmental Sustainability: (Vol-II) Approaches to Tackle Pollution for Cleaner and Greener Society* (Elsevier, The Netherlands); and *New Trends in Removal of Heavy Metals from Industrial Wastewater* (Elsevier, USA)—and the author of two books, *Recent Advances in Distillery Waste Management for Environmental Safety* (CRC Press, USA) and *Microbe-Assisted Phytoremediation of Environmental Pollutants: Recent Advances and Challenges* (CRC Press, USA). He has presented several papers relevant to his research areas at national and international conferences. He is an active member of various scientific societies, including the Association of Microbiologists of India (AMI), the Biotech Research Society (BRSI), and the Indian Science Congress Association (ISCA). He is secretary-general of the Society for Green Environment, Delhi, India. He can be reached at vineet.way18@gmail.com or drvineet.micro@gmail.com.

 **Gaurav Saxena** has been an Assistant Professor of Microbiology at the Department of Microbiology, School of Life and Allied Sciences (SLAS), and a member of the Proctorial Board at Baba Farid Institute of Technology, Dehradun, Uttarakhand, India, since 2019. He was born in 1989 and completed school education at Government Schools at Shahjahanpur, Uttar Pradesh, India. Dr. Saxena received his BSc (2010) in Industrial Microbiology, Zoology, Botany, and Chemistry from Hemwati Nandan Bahuguna Garhwal (Central) University (HNBGU), Srinagar (Garhwal), Uttarakhand, India. He received his MSc (2013) in Environmental Microbiology from the Babasaheb Bhimrao Ambedkar (Central) University, Lucknow, UP, India, where his interest in environmental issues began with work on environmental remediation using microorganisms. At the same university, Dr. Saxena earned his PhD (2020) in Environmental Microbiology (specializing in microbial bioremediation and metagenomics research). He then went to Jawaharlal Nehru University (JNU), New Delhi (India), where he was trained on a Department of Biotechnology (DBT)-sponsored research project for the production of biofuel from oleaginous bacteria under the supervision of Prof. Indu Shekhar Thakur. He is an active young researcher in the areas of Environmental Microbiology, Environmental Toxicology, Bio/Phytoremediation, Metagenomics, and Biological Wastewater Treatment. Dr. Saxena teaches courses in General Microbiology, Molecular Biology, and Microbial Genetics, Environmental Microbiology, and Laboratory Techniques in Microbiology and Molecular Biology. Currently, Dr. Saxena is engaged in sustainable environmental remediation technologies for environmental safety. He has been qualified (2016) in the National Eligibility Test (NET) and received the Junior Research Fellowship (JRF) of the Department of Science and Technology (DST), Government of India (GOI) New Delhi, India. Dr. Saxena has been awarded prestigious awards like *Young Scientist Award*, *Young Environmentalist Award*, and *Young Achiever Award* in recognition of his scientific research. He is the editor of five books—*Bioremediation for Environmental Sustainability: Toxicity, Mechanisms of Contaminants Degradation, Detoxification and Challenges* (Elsevier, USA); *Bioremediation for Environmental Sustainability: Approaches to Tackle Pollution for Cleaner and Greener Society* (Elsevier, The Netherlands); *Bioremediation of Industrial Pollutants* (Write and Print Publication, India); *Bioremediation of Industrial Waste for Environmental Safety: Industrial Waste and Its Management* (Springer Nature, Singapore), and *Bioremediation of Industrial Waste for Environmental Safety: Biological Agents and Methods for Industrial Waste Management* (Springer Nature, Singapore)—and author of one book, *Microbe-Assisted Phytoremediation of Environmental Pollutants: Recent Advances and Challenges* (CRC Press, USA). He is also on the editorial board of the *Frontiers in Microbiology* (a Swiss Journal) and serving as a review editor. Dr. Saxena's academic contributions include several scientific papers, reviews and book reviews, conference papers, book chapters, popular science articles, weblished articles, and, short article in reputed national and international journals

(Springer Nature, Elsevier, and International Water Association (IWA)), conference proceedings, books (Springer Nature, CRC Press, and Elsevier), science magazines, websites, and newspapers, respectively. Currently, Dr. Saxena is serving as a potential reviewer for several international and national research journals in his thrust areas. He is a member of the Association of Microbiologists of India (AMI) and the Indian Science Congress Association (ISCA), American Chemical Society (ACS), and British Ecological Society (BES). He is a nature lover and has a dedicated cause for environmental protection. He can be reached at gaurav10saxena@gmail.com.

1 Microbe-Assisted Phytoremediation

A Promising Technology for Remediation of Environmental Pollutants

1.1 INTRODUCTION

Rampant industrialization, urbanization, and anthropogenic activities are associated with a continuous build-up of environmental contaminants and pollution, posing considerable risk to humans and environmental health, leading to epidemics of cancers, lung diseases, and other degenerative diseases (Chandra et al. 2018a,b). Among the different sources of environmental contaminants, industries are considered the major source of soil, water, and air pollution. To obtain good-quality products within a short period of time, industries generally use poorly biodegradable or non-biodegradable chemicals and subsequently generate a huge quantity of hazardous and toxic waste. Industrial waste, containing a mixture of numerous organic and inorganic pollutants, is generally dumped on land and/or discharged into water bodies, and thus they become a large source of environmental pollution and health hazards. In wealthy industrialized countries, contamination is often highly limited to a small area, and the pressure to use contaminated land and water for agricultural food production or human consumption, respectively, is minimal. However, soil and water contamination is increasingly recognized as dramatic in large parts of the developing world. A large array of methods based on not only physical and chemical but also biological means have been available for the remediation of contaminated soil and water for decades; however, these methods are financially expensive and energy intensive, cause secondary air or groundwater pollution, and can destroy waste-degrading microbial communities of soil, and for many refractory pollutants, no feasible technologies are yet available. Therefore, environmental preservation requires the development of sustainable approaches that promise thorough, economical, and eco-friendly ways to make them safe for human habitation and consumption and to protect the functions of the life-supporting ecosystem. To safeguard both humans and the environment from the adverse consequences of organic and inorganic pollutants, novel approaches must be designed, and

1

phytoremediation is one such approach. Phytoremediation is an ideal approach for the treatment and/or elimination of toxic organic and inorganic pollutants from the contaminated environment or to render them harmless (Khan et al. 2014a; Chandra et al. 2015). It is an emerging green approach where plants are grown in contaminated soil, sediment, and surface and groundwater to increase the decomposition or removal rate of inorganic and organic pollutants *in planta* as well as *ex planta*. Plants possess extremely efficient root systems that acquire and concentrate nutrients from the contaminated matrices as well as numerous metabolic activities, all of which are ultimately powered by photosynthesis, but the major constraint of this technology is that even plants that are tolerant to the presence of contaminants often remain relatively small, due to the toxicity of the pollutants that they are accumulating from contaminated matrices or the toxic end-products of their degradation (Kumar and Chandra 2018a). Besides, it is a time-consuming remediation technology, with slow degradation and limited uptake of organic and inorganic contaminants from the contaminated matrices. The toxicity of organic and inorganic contaminants on plants can be reduced by using a microbe in association with the plant (Weyens et al. 2009a,b; Glick 2012; Khan et al. 2014a,b). The combinatorial systems of plants and their associated microbes have been shown to contribute to biodegradation and detoxification of organic and inorganic compounds in polluted soil and water and could have potential for improving the phytoremediation efficacy of plants (Rajkumar et al. 2012; Redfern and Gunsch 2016; Fatima et al. 2018). Microbe-assisted phytoremediation is a promising, inexpensive, and eco-friendly rehabilitation approach that uses a broad range of plants and their associated microbes for remediating pollutants present in different environmental matrices (Juwarkar and Singh 2010; Nanekar et al. 2015; Ali et al. 2017; García-Sánchez et al. 2018). They can metabolize, detoxify and/or biotransform many refractory pollutants either to obtain carbon and/or energy for their growth or as co-substrates, thus converting them to simpler products such as carbon dioxide (CO_2) and water (H_2O).

This chapter presents an overview of the environmental pollution, toxicity profile, and health hazards of numerous organic and inorganic pollutants discharged from various industries and their management approaches using physical, chemical, and biological means. We also discuss the merits and demerits of physico-chemical and biological methods used for the remediation of organic and inorganic pollutants from contaminated sites. Further, we also describe various processes of phytoremediation, with special emphasis on microbe-assisted phytoremediation as a cleanup technique for remediation of contaminated sites and provide a concise discussion of how microbes could be exploited to enhance the phytoremediation efficacy of plants in contaminated environments.

1.2 POLLUTANTS AND THEIR FATE IN THE ENVIRONMENT

Solid and liquid waste discharged after various industrial operations is considered a major source of hazardous, toxic, and refractory pollutants in the environment (Calheiros et al. 2009; Chandra and Kumar 2015a,b, 2017a,b; Enazy et al. 2017; Mesa et al. 2017). In general, there are two types of environmental pollutants (i) organic pollutants and (ii) inorganic pollutants.

1.2.1 ORGANIC POLLUTANTS

Organic pollutants may include various compounds, such as petroleum hydrocarbons (i.e., benzene, pyrene, toluene, and xylene), chlorinated solvents (i.e., polychlorinated biphenyls [PCBs], trichloroethylene [TCE]), polycyclic aromatic hydrocarbons (PAHs), persistent organic pollutants (POPs) (i.e., aldrin, chlordane, dichlorodiphenyltrichloroethane [DDT], dieldrin, endrin, heptachlor, hexachlorobenzene, mirex, toxaphene, chlordecone, lindane, hexachlorobenzene, pentachlorobenzene, α-hexachlorocyclohexane, β-hexachlorocyclohexane, perfluorooctane sulfonic acid and its salts, perfluorooctane sulfonyl fluoride, and tetrabromodiphenyl), endocrine-disrupting chemicals (EDCs) (i.e., disulfide; o-phenylphenol; tetrabrominated diphenyl ether; 4-chloro-3-methylphenol; 2,4-dichlorophenol; resorcinol; 4-nitrotoluene; 2,2'-bis(4-(2,3-epoxypropoxy) phenyl) propane; 4-octylphenol; estrone [E1]; 17α-ethinylestradiol [EE2]; and 17β-estradiol [βE2]), azo dyes, melanoidins, organophosphorus compounds (i.e., glyphosate, chlorpyrifos, parathion, monocrotophos, dicrotophos, diazinon, dimethoate, fenitrothion), volatile organic carbons (VOCs), and explosives (i.e., nitroglycerine [NG]; 2,4,6-trinitrotoluene [TNT]; hexahydro-1,3,5-trinitro-1,3,5-triazine [Royal Demolition Explosive—RDX]; octahydro-1,3,5,7-tetranitro-1,3,5-tetrazocine [HMX]; and pentaerythritol tetranitrate [PETN]). One of the primary concerns about the environment's exposure to organic compounds is their potential to contaminate aquatic and terrestrial ecosystems and consequently posea risk to humans and other organisms associated with the food chain of the aquatic and terrestrial eco-biota (Ying and Rai 2003; Adeola 2004; Katrin et al. 2005; Singh and Walker 2006; Taioli et al. 2007; Carpenter 2011; Manzetti 2013; Igbinosa et al. 2013; Chandra and Kumar 2015a, 2017a,b; Faroon and Ruiz 2016; Hussein and Scholz 2017; Carnevali et al. 2018; Kumar and Chandra 2018a; Yaseen and Scholz 2019; Kumar et al. 2020).

1.2.2 INORGANIC POLLUTANTS

Inorganic pollutants may include non-biodegradable heavy metals, metalloids, and nonradioactive or radioactive nuclides such as uranium (U), vanadium (V), wolfram (W), strontium (Sr), and cesium (Cs). Inorganic pollutants also include various nutrients like ammonia, chloride, sodium, nitrate, and phosphate (Chandra and Kumar 2015a,b, 2017c). Heavy metals (HMs) and metalloids are the main groups of inorganic contaminants generated through natural sources such as weathering of minerals, erosion and volcanic activities, and forest fires, and particles released by vegetation and/or anthropogenic sources include human activities such as mining, smelting, ore processing, irrigation by sewage water, injudicious use of chemical fertilizers and pesticides, pile-up of municipal waste, automobile exhaust, electroplating, leather tanning, textiles and dyeing, distilleries, pulp and paper industries, and other industrial and domestic activities that pour directly or indirectly into the environment, and a considerably large area of land and water is contaminated with them (Wuana and Okieimen 2011; Tchounwou et al. 2012; Chandra et al. 2018a,b). The term "HMs" refers to any metallic element that has a relatively high atomic weight (>20) and high density (>4 g/cm^3, or 5 times or more than water) and is toxic to organisms even at very low concentrations (Ali et al. 2013, 2019). HMs include mercury (Hg), cadmium (Cd), chromium (Cr), zinc (Zn), aluminum (Al), cobalt (Co), copper (Cu), iron (Fe), molybdenum (Mo), manganese (Mn), lead (Pb), nickel (Ni), magnesium (Mg), selenium (Se), and silver (Ag). On the other hand, a metalloid is a chemical element that has properties between those of metals and nonmetal elements. The six commonly recognized metalloids are boron (B), silicon (Si), germanium (Ge), arsenic (As), antimony (Sb), and tellurium (Te). Out of six metalloids, two metalloids, Sb and As, are highly toxic in nature. Most HMs and metalloids are electronegative in nature and often non-biodegradable and thus persist in the environment, causing serious soil and water pollution and health threats in living beings (Ali et al. 2019). HMs and metalloids such as As, Cd, Cr, Pb, and Hg are listed as priority pollutants by the United States Environmental Protection Agency (USEPA) (Tchounwou et al. 2012; Ali et al. 2019). These elements can cause toxicity to humans and other forms of life even at low concentrations; their deleterious effects on living beings and sources in the environment are quite evident from Table 1.1. However, living organisms require varying amounts of HMs for physiological functions of living tissues, and they regulate many biochemical processes, but higher concentrations can notoriously lead to health problems and severe poisoning. It has been reported that HMs such as Co, Cu, Fe, Cr, Se, Mo, Mn, and Zn are required in minute quantities for plant growth and are classified as essential micronutrients that are required for various biochemical and physiological functions. Other HMs and metalloids that are commonly found as contaminants and are nonessential for plants include As, Cd, Cr, Hg, Ni, Pb, Se, U, V, and W. However, excessive amounts of these elements can become harmful to plants, even at low concentrations. Because of their high solubility in aquatic environments, HM scan be absorbed by living organisms (Jaishankar et al. 2014). Once they enter the food chain, large concentrations of HMs may accumulate in the human body. If the metals are ingested beyond the permitted concentrations, they can cause serious health disorders (Järup 2003). However, Fe, Co, Cu, Mn, and Zn are required for physiological functions of living tissues by humans. Additionally, Hg and Pb are toxic metals that have no known vital or beneficial effect on organisms, and their accumulation over time in the bodies of animals and humans can cause serious illness (Jaishankar et al. 2014). Organic and inorganic environmental pollutants are highly toxic and hazardous in nature and cause serious negative consequences, such as damage to ecosystems and agricultural productivity, deterioration of the food chain, contamination of water resources, economic damage, and, finally, serious human and animal health problems (Lambolez et al. 1994; Shao et al. 2014; Saxena et al. 2017; Kumar and Sharma 2019). Thus, the recovery of sites contaminated with such compounds is one of the major challenge for environmentalists, scientist, and researchers. The toxic effects caused by organic and inorganic pollutants in living beings (animals and plants) and brief accounts of their toxicity are presented in the Table 1.1.

TABLE 1.1

Different Heavy Metals, Their Permissible Levels, and Their Effects on Human and Plant Health and Sources

Heavy Metal	Standard Regulatory Limit		Toxic Effects on the Various Body Parts of Human	Major Sources in Environment	Toxic Effects on Plants
	Soil	Water			
As	0.02	–	Bronchitis; carcinogenic dermatitis; liver tumors; gastrointestinal damage; conjunctivitis; dermatitis; perforation of nasal septum; respiration cancer; peripheral neuropathy; skin cancer; chronic diseases; pre-malignant skin lesions; lung, bladder, liver, and kidney damage; cardiovascular disease; edema; weakness; hypertension; respiratory problems; neurological deficits; oxidative stress	Volcanic eruptions, soil erosion, pesticides, fungicides, metal smelters, coal fumes, wood preservatives, semiconductors, petroleum refining, animal feed additives, coal power plants, herbicides, mining and smelting	–
Cd	0.06	–	Kidney damage; bronchitis; gastrointestinal disorders; bone marrow; cancer; weight loss; carcinogenic, mutagenic, and teratogenic; endocrine disruptor; interferes with calcium regulation in biological systems; causes renal failure, coughing, emphysema, headache, hypertension, prostate cancer, lymphocytosis, microcytic hypochromic anemia, testicular atrophy, osteoporosis, and fractures; pulmonary and gastrointestinal irritant	Welding, electroplating, pesticides, fertilizers, batteries, nuclear fission plants, geogenic sources, metal smelting and refining, fossil fuel burning, application of phosphate fertilizers, sewage sludge	–

(Continued)

TABLE 1.1 (Continued)
Different Heavy Metals, Their Permissible Levels, and Their Effects on Human and Plant Health and Sources

Heavy Metal	Standard Regulatory Limit		Toxic Effects on the Various Body Parts of Human	Major Sources in Environment	Toxic Effects on Plants
	Soil	Water			
Cr	0.01	0.01	Bronchopneumonia; dizziness; chronic bronchitis; diarrhea; emphysema; headache; irritation of the skin; itching of the respiratory tract; liver diseases; lung cancer; nausea; renal failure; reproductive toxicity; vomiting; allergic dermatitis; lung tumors; human carcinogens; dermal, respiratory, cardiovascular, gastrointestinal, hematological, hepatic, renal, neurological, developmental, reproductive, immunological, genotoxic, mutagenic and carcinogenic effects	Burning of oil and coal, petroleum from ferrochromate refractory material, pigment oxidants, catalysts, chromium steel, fertilizers, oil well drilling, metal plating, tanneries, steel industry, mining, cement, paper, rubber, metal alloy paints, electroplating industry, sludge, solid waste, wood preservation, chemical production, pulp and paper production	Reduction in root growth, leaf chlorosis, inhibition of seed germination, depressed biomass, chlorosis and necrosis in plants
Cu	3.0	—	Irritation of nose, mouth, eyes; headache; stomachache; dizziness; brain and kidney damage; liver cirrhosis and chronic anemia; stomach and intestinal irritation; shyness; tremors; memory problems; irritability; and changes in vision or hearing; lung damage; vomiting; diarrhea; nausea; skin rashes	Brass manufacture; electronics; electrical pipes; additive for antifungals; electroplating industry; smelting and refining; mining; biosolids; volcanic eruptions; forest fires; emissions from industries producing caustic soda, coal, or peat; woodburning	—

(Continued)

TABLE 1.1 (Continued)
Different Heavy Metals, Their Permissible Levels, and Their Effects on Human and Plant Health and Sources

Heavy Metal	Standard Regulatory Limit		Toxic Effects on the Various Body Parts of Human	Major Sources in Environment	Toxic Effects on Plants
	Soil	Water			
Hg	0.01	—	Dermatitis; nervous system and kidney damage; anorexia; protoplasm poisoning; severe muscle pain; blindness; deafness; decreased rate of fertility; dementia; dizziness; dysphasia; gastrointestinal irritation; gingivitis; kidney problems; loss of memory; pulmonary edema; reduced immunity; sclerosis; damage to brain, kidney, and developing fetus; corrosive to skin, eyes and muscle membranes	Pesticides, batteries, paper and leather industry, thermometers, electronics, pharmaceuticals, electrical industry (switches, thermostats, batteries), dentistry (dental amalgams), municipal wastewater discharge, mining, incineration, discharge of industrial wastewater	—
Pb	0.1	—	Kidney and gastrointestinal (GIT) damage, sterility, anemia, muscle and joint pains, hypertension, impaired development, anorexia, chronic nephropathy, damage to neurons, hyperactivity, insomnia, renal failure, mental retardation in children	Mining and smelting of metalliferous ores, burning of leaded gasoline, municipal sewage, industrial waste enriched in Pb, pesticides, smoking, batteries, water pipes, automobile emissions, mining, burning of coal, lamps, house paint, plumbing pipes, pewter pitchers, storage batteries, toys, faucets	Production of reactive oxygen species (ROS), causing lipid membrane damage that ultimately leads to damage of chlorophyll and photosynthetic processes and suppresses the overall growth of the plant, reducing biomass; strong inhibition of plant growth

(Continued)

TABLE 1.1 (*Continued*)

Different Heavy Metals, Their Permissible Levels, and Their Effects on Human and Plant Health and Sources

Heavy Metal	Standard Regulatory Limit		Toxic Effects on the Various Body Parts of Human	Major Sources in Environment	Toxic Effects on Plants
	Soil	Water			
Zn	5.0	–	Nervous membrane and skin damage; short-term illness called metal fume fever; restlessness; over dosage causes ataxia, depression, gastrointestinal irritation, hematuria, impotence, kidney and liver failure, lethargy, macular degeneration prostate cancer, seizures, vomiting	Electroplating industry, smelting, refining, mining, biosolids, refineries, brass manufacture, metal plating, plumbing	–
Ni	3.0	–	Causes chronic bronchitis; allergic dermatitis known as nickel itch; inhalation can cause cancer of the lungs, nose, sinuses, throat, and stomach; immunotoxic, neurotoxic, genotoxic, reproductive toxic, pulmonary toxic, nephrotoxic, and hepatotoxic; causes hair loss	Steel industry, mining, magnetic industry, volcanic eruptions, landfill, forest fires, bubble bursting and gas exchange in ocean, weathering of soils and geological materials	Decrease in chlorophyll content and stomatal conductance, affected Calvin cycle and CO_2 fixation, reduction in plant nutrient acquisition, decrease in shoot yield, chlorosis

1.3 REMEDIATION APPROACHES FOR ENVIRONMENTAL POLLUTANTS

1.3.1 PHYSICO-CHEMICAL APPROACHES

A large number of various physico-chemical methods, such as coagulation/flocculation, electrocoagulation, thermolysis, membrane filtration, advanced oxidation processes (AOPs), radiation excavation, landfills, thermal treatments, leaching, and electro-reclamation are being used for remediation of environmental pollutants from contaminated water, wastewater, sediment, and soil (Sharma et al. 2018). These techniques are rapid but inadequate and costly and generate secondary pollution. They only change the problem from one form to another and do not completely remediate the pollutants from contaminated matrices (Hashim et al. 2011; Azubuike et al. 2016).

1.3.2 BIOLOGICAL APPROACHES

Biological approaches have been recognized as eco-friendly and the most effective methods applied for the remediation of organic and inorganic pollutants from the contaminated site for environmental safety and human health protection (Ye et al. 2017). Biological approaches generally include the following methods.

1.3.2.1 Bioremediation

Bioremediation is an eco-friendly and sustainable method that employs many different organisms and works in parallel or series of the sequence to vitiate and/or detoxify toxic contaminants (Kumar et al. 2018). In other words, it can be stated as the speeding up of the normal metabolic process, where microorganisms (i.e., bacteria, actinomycetes, fungi, yeast, and cyanobacteria) and green plants (a process termed phytoremediation) or their enzymes disintegrate or transform toxic contaminants into CO_2, H_2O, inorganic salts, and other by-products (metabolites) that are less toxic than the parental compounds (USEPA 2000). Employing microbes for degradation and detoxification of pollutants is now being increasingly applied as the technology of preference for cleanup or to restore contaminated sites to a sustainable environment. On the basis of transportation and eradication of pollutants from contaminated sites, bioremediation technology can be grouped into two categories: (i) *in situ* bioremediation and (ii) *ex-situ* bioremediation (USEPA 2006, 2012).

1.3.2.1.1 In Situ *Bioremediation*

In situ bioremediation is the process performed at the original contaminated site. In this technique, oxygen/air and nutrients are added to the contaminated site in order to accelerate the growth of pollutant-degrading indigenous microbes and escalate the rate of biodegradation (USEPA 2006). *In situ* bioremediation technology can be divided into two categories: (i) intrinsic *in situ* bioremediation and (ii) engineered *in situ* bioremediation. Intrinsic *in situ* bioremediation (also known as natural attenuation [NA] or passive bioremediation) is a degradation process of organic pollutants persisting in the contaminated soil by metabolic activities of indigenous microbes, without utilizing any artificial augmentation (USEPA 1999). On the other hand, engineered *in situ* bioremediation, also called *in situ* bioremediation, is the introduction of indigenous microbes to contaminated sites to accelerate the biodegradation process by developing or enhancing the conducive physico-chemical conditions of an environment (USEPA 2000, 2006). In the engineered *in situ* bioremediation process, nitrogen, oxygen, and phosphorus are disseminated through the subsurface via an instilment or extraction well in order to stimulate the growth and metabolism of indigenous microbes. Microbes using oxygen as an electron acceptor convert it to H_2O as they disintegrate the toxic pollutants. Examples of engineered *in situ* bioremediation-based techniques are bioventing, biosparging, bioslurping, biostimulation, and bioaugmentation (Figure 1.1; USEPA 2006).

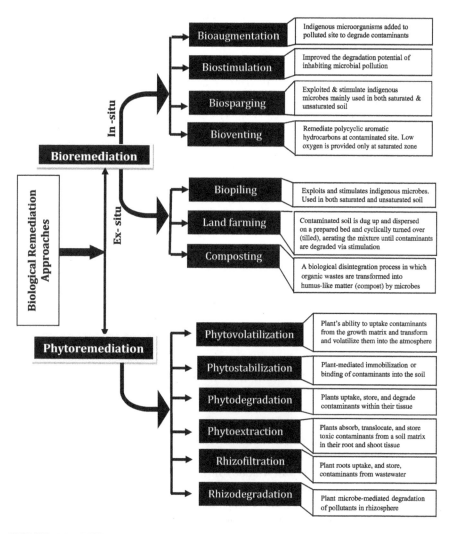

FIGURE 1.1   Different strategies of plants and microbes for remediation of organic and inorganic pollutants.

1.3.2.1.2   Ex-Situ *Bioremediation*

Ex-situ bioremediation methods involve the digging of contaminated matrices like soil, sediment, or sludge and pushing of groundwater to facilitate microbial degradation of pollutants (USEPA 2006). *Ex-situ* bioremediation can take place in two ways: (i) slurry-phase bioremediation and (ii) solid-phase bioremediation. Slurry-phase bioremediation is a biological procedure where the contaminated matrices are blended with H_2O and other chemicals within a bioreactor, and blending conditions are maintained to promote the biodegradation rate of water-soluble and soil-connected pollutants present in the water slurry of the contaminated matrices and biomass of indigenous microbes (Kumar and Chandra 2020a). Afterward, O_2 and nutrients are supplemented to the reactor to establish the ideal environmental conditions for microbes to degrade the specific pollutants. Once the process is completed, the H_2O is withdrawn from the reactor, and the contamination-free media is assessed and replenished in the environment. On the other hand, solid-phase bioremediation is a method that treats the contaminated soil within an above-ground treatment area. Conditions within the treatment areas are monitored in order to ensure optimum treatment is taking place. This bioremediation process is based on the mechanical breakdown of polluted soil by abrasion and an intensive blend of the components in an enclosed vessel. This confirms that the nutrients, microbes, pollutants, O_2, and H_2O are in permanent contact. Solid-phase soil treatments include soil biopiles, land farming, and composting practices for detoxification and disintegrating hazardous toxic contaminants (Figure 1.1).

1.3.2.1.3   Electrobioremediation

Electrobioremediation (EKB) is a generic name for a hybrid technology coupling bioremediation to electrokinetics (EK) for the treatment of soil contaminated with petroleum and its components (Chilingar et al. 1997; Zhang et al. 2017). It comprises a large group of engineered cleanup methods that apply electrokinetic phenomena for acceleration and direct transport of contaminants (or their derivatives), nutrients, electron acceptors, and catabolically active microorganisms in a subsurface. The fundamental idea of applying electrokinetics to bioremediation is to stimulate contaminant biodegradation by effective homogenization of microorganisms and contaminants without further extensive mechanical treatment of the subsurface matrix (Yan and Reible 2015; Wick et al. 2007). The EKB process is highly efficient during the initial stages of treatment owing to the stimulation of bacterial activity by the weak electric field. However, electrokinetic remediation changes the soil properties, including the soil pH, moisture, and microbial biomass (Wang et al. 2016).

1.3.2.2   Phytoremediation: Plants Take up the Challenge

Phytoremediation, also referred to as botanical bioremediation or green remediation, is widely considered a cost-effective, aesthetically pleasing, solar-driven and eco-friendly technique to eliminate hazardous and toxic HMs and organic pollutants present in soil, sludge, sediment, water, or wastewater (Cunningham and Berti 1993; Salt et al. 1995; Ali et al. 2013; Chandra et al. 2018a,b). The term "phytoremediation," derived from the Greek phyto (plant), and Latin remedium (cure, remedy, or heal), first appeared in the pertinent literature in the early 1990s, in a paper by Cunningham and Berti (1993). Yet, in the scientific domain, the concept of using plants to mitigate pollution was first proposed a decade before (Chaney 1983). According to Cunningham and Berti (1993) and Cunningham et al. (1995), phytoremediation is defined as the engineered use of green plants, including grasses, forbs, and woody species, to remove, contain, or render harmless environmental contaminants present in soil, sludge, sediment, water, or wastewater. Depending on the type of contaminants, the site conditions, and the level of cleanup required, phytoremediation technology is currently divided into a number of processes, namely phytoextraction, phytofiltration, phytostabilization, phytovolatilization, and phytodegradation, with each process having a different mechanism of action for remediation of environmental pollutants from the contaminated site (Ali et al. 2013; Chandra and Kumar 2018). The different processes of phytoremediation and their mechanisms are illustrated and described in Figure 1.2 and Table 1.2, respectively.

1.3.2.2.1   Phytoextraction

Phytoextraction refers to the use of pollutants accumulating plants to absorb, translocate, and store metals or organic contaminants from soil, sediments, sludge,

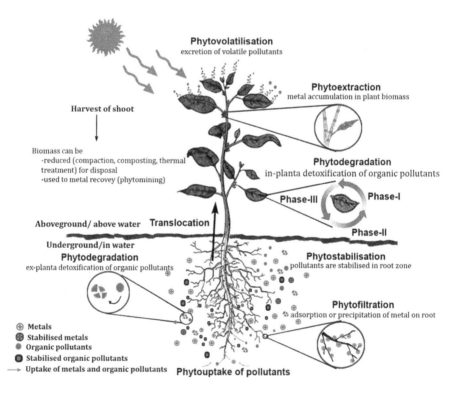

FIGURE 1.2   Different strategies of plant for phytoremediation of environmental pollutants from contaminated sites. (Reprinted with permission from Chandra R, Kumar V. 2018. In: Chandra R, Dubey NK, Kumar V (eds) *Phytoremediation of Environmental Pollutants*. Boca Raton, FL: CRC Press.)

TABLE 1.2

Summary of Phytoremediation Techniques and Their Mechanisms in Various Contaminated Media

Phytoremediation Technique	Description	Medium Treated	Action Mechanism	Applicability
Phytoextraction/phytoaccumulation/phytoabsorption/phytosequestration	Accumulation of contaminants in plant shoots with subsequent removal from contaminated soil	Soil, wastewater, sludge	Translocates and concentrates contaminants from the soil via plant roots into harvestable plant parts	Inorganic pollutants, e.g., metals
Phytofiltration	Utilizes plants with extensive root systems for removal of pollutants from contaminated water or wastewater	Surface water, wastewater	Sequestration of pollutants by plant root from contaminated water or wastewater	Inorganic/organic pollutants
Phytostabilization/phytimmobilization/*in situ* or in-place inactivation	Immobilizes pollutants and reduces their bioavailability	Groundwater, soil, mine tailings	Mobility of pollutants becomes limited by the plant rootsand/or root exudates in rhizosphere	Inorganic pollutants
Phytodegradation/phytotransformation	Breakdown of organic contaminants through plant metabolic activities or plant enzymes	Surface and groundwater within the rhizosphere and soil	Absorption and uptake of pollutants by root system, resulting in metabolic or enzymatic transformation within or external to plants	Organic pollutants or xenobiotic compounds
Phytovolatilization	Pollutants converted into volatile form or gas phase and their subsequent liberationto the atmosphere through transpiration	Soil, groundwater	Modification of pollutants during vascular translocation from roots to leaves	Inorganic/organic pollutants
Rhizoremediation/phytostimulation, plant-assisted bioremediation/microbe-assisted phytoremediation/rhizosphere bioremediation	Degrades pollutants by soil-dwelling microbes in rhizosphere due to simulation of microbial activity by plant secretions	Soil, wastewater, sludge	Secretion of root exudates or enzymes in rhizosphere and subsequent microbial degradation of contaminants	Organic pollutants or xenobiotic compounds

water, and/or wastewater in their root and shoot tissues (Salt et al. 1998; Garbiscu and Alkorta 2001; Chandra and Kumar 2017c). The term phytoextraction is mostly used to refer to removal of metal from contaminated matrices. Generally, the ideal plants to be used in phytoextraction should have the following characteristics: (i) toleration of high levels of metal concentration; (ii) fast growth rate and high biomass production; (iii) ability to accumulate high level of metals in their harvestable parts; (iv) widely distributed and with a deep root system; (v) resistance to disease and pests and unattractive to animals; (vi) easy cultivation, harvesting, and processing; and (vii) repulsive to herbivores, to avoid food chain contamination (Cunningham et al. 1997; Chandra and Kumar 2018). Phytoextraction can be performed by three means (i) phytoextraction by trees, (ii) phytoextraction by crops, and (iii) phytoextraction by grasses. Each has its own advantages and disadvantages (Raskin et al. 1997; McGrath and Zhao 2003).

1.3.2.2.1.1 Evaluation of Phytoextraction Potential of Plants Two important parameters are generally used to assess the potential of a given plant species suitable for phytoextraction: the bioconcentration factor (BCF) and the translocation factor (TF) (Yoon et al. 2006). BCF is defined as the ratio of the metal concentration in the root of the plant to that in the soil/sludge. The TF denotes the plant's ability to translocate metals from the root to its harvestable parts.

$$BCF = C_{root}/C_{soil/sludge}$$

$$TF = C_{shoot}/C_{root}$$

where C_{root}, $C_{soil/sludge}$, and C_{shoot} are the concentrations of metal in the root, soil/sludge, and shoot, respectively.

Thus, the evaluation and selection of plants for phytoremediation purposes entirely depends on the BCF and TF. A plant with both BCF and TF values of >1 has the potential for use in phytoextraction (Yoon et al. 2006). Additionally, a plant with a BCF of >1 and TF of <1 has potential for phytostabilization.

1.3.2.2.1.2 Plants That Grow on Metal-Contaminated Sites and Their Tolerance Mechanisms According to three basic survival strategies on metalliferous and contaminated soil, plants have been categorized as metal excluders, indicators, and accumulators (Baker 1981; Baker and Brooks 1989a; Baker and Walker 1989b). *Metal excluders*—effectively prevent metal from entering their aerial parts over a broad range of metal concentrations in the soil; however, they can still contain large amounts of metals in their roots. *Salix* (willow) species are examples of metal excluders. *Metal indicators*—accumulate metals in above-ground tissues to levels similar to those in the surrounding soil. A recent example demonstrates the potential use of *Eucalyptus* trees, which translocate Au from deep mineral deposits to their aerial tissues, to identify Au deposits (Lintern et al. 2013). Metal *accumulators* are usually referred to as hyperaccumulators that concentrate metals in their above-ground tissues to levels far exceeding those present in the soil or in non-accumulating species growing nearby. Pioneering work by the late Professor Robert Brooks at Massey University,

Palmerston North, New Zealand, popularized the study of plants that accumulate inordinate amounts of heavy metals. Chaney was the first to suggest using these hyperaccumulators for the phytoremediation of metal polluted sites. According to Baker and Brooks (1989a), hyperaccumulators are defined as plants that take up and accumulate >1 mg kg^{-1} dry weight of Au; >100 mg kg^{-1} dry weight of Cd; >1000 mg kg^{-1} dry weight of Cu, Co, Cr, Ni, or Pb; or >10 000 mg kg^{-1} dry weight of Mn or Zn in their shoots when grown on metalliferous soil. Metal concentrations in the shoots of accumulating plants can be 100- to 1000-fold higher than in non-accumulating plants. Apart from these rather arbitrary criteria, hyperaccumulator plants usually have a shoot-to-root metal concentration ratio of >1, whereas non-hyperaccumulator plants generally have higher metal concentrations in roots than in shoots (Baker et al. 1994; Shen et al. 1997). Up to now, approximately 500 plant species, including trees, vegetable crops, grasses, and weeds from at least 45 plant families have been reported to be hyperaccumulators for various metals, and most of them are Ni hyperaccumulators occurring in ultramafic areas all over the world (Rascio and Navari-Izzo 2011). Some important families are *Brassicaceae, Euphorbiaceae, Asteraceae, Fabaceae, Lamiaceae,* and *Scrophulariaceae.* Although metal hyperaccumulator plants have ideal properties for phytoextraction of metals from contaminated sites, most of these plants produce little biomass and are thus primarily used as model organisms for research purposes (Suman et al. 2018).

Commonly, plants grown on metalliferous soil show two types of tolerance mechanisms against toxic metals. The first is constitutive and the second is adaptive. Adaptive is the most common mechanism acquired by metal-tolerant plants and includes immobilization, plasma membrane exclusion, synthesis of specific heavy metal transporters, induction of stress proteins, restriction of uptake and transport, chelation of HMs by particular ligands/peptides and transportation of these complexes to above-ground tissues where they are sequestered into vacuoles, and biosynthesis of polyamines and signaling molecules such as salicylic acid and nitric oxide (Hossain et al. 2012). Among all adaptive mechanisms, phytochelatin of HMs is one of the most important mechanisms, in which root exudates chelate HMs and these chelated metals bind easily to the wall of the plant cell and enter into the cell. Inside the cells, a high-affinity peptide compound is produced that binds HMs and controls their cytoplasmic concentration by transporting them across the tonoplast and to their subsequent sequestration in the vacuole (Yang et al. 2005; Benyó et al. 2016). Two main classes of chelating peptides that play an important role in the detoxification of metals are phytochelatins (PCs) and metallothioneins (MTs), and their synthesis in the plant is induced by exposure of root cells to HMs (Yang et al. 2005; Rascio and Navari-Izzo 2011).

PCs are cysteine-rich polypeptides enzymatically synthesize from glutathione (GSH) in response to HM stress with general structure (c-Glu-Cys)n-Gly that chelates HMs in the cytosol of plant cells and subsequently play a major role in HM detoxification and tolerance in plants. Besides plants, PCs have been reported in yeasts, protists, and nematodes, but there is no evidence for their presence in humans or other mammals. PCs form ligand complexes with HMs, which are then sequestered into the vacuole. On the other hand, MTs are sulfur-rich low-molecular-weight peptides of 60–80 amino acids that contain 9–16 cysteine residues by which plants combat abiotic

HM stresses through binding of metal ions in their thiol groups of cysteine residues. In addition, MTs are involved in the metal tolerance or homeostasis and scavenging of reactive oxygen species and virulence of fungi in plants. In addition to plants, MTs are found in animals and some prokaryotes. The role of MTs in HM detoxification has also been ascertained in animals and microbes. In animals, MTs have a protective role against Cd toxicity, but in plants, this role is mostly fulfilled by MTs and PCs together (Cobbett and Goldsbrough 2002). Besides peptides (phytochelators), plants also produce biosurfactants, which consist of glycosides, and polyphenols, generally found in many plant species, play an important role in metal removal because of acidic characteristics and the presence of numerous functional groups. Plants play a crucial role in the remediation of heavy metals; to understand the mechanism clearly underlying their genetic basis can be an important aspect regulated by an inter-related network of physiological and molecular mechanisms (Yang et al. 2005). Various species of plants possess different kinds of mechanisms to tolerate heavy metals, although within one species, more than one mechanism could be in operation.

1.3.2.2.2   Phytostabilization

Phytostabilization is another strategy where plants, either physically or by the action of root exudates, sequester contaminants, mainly HMs, in the soil, making them less bioavailable. In this process, certain plant species immobilize contaminants in the soil through adsorption on roots, accumulation by the roots, or precipitation within the root zone and physical stabilization of the soil, reducing the risk to human health and the environment.

1.3.2.2.3   Phytovolatilization

The release of volatile pollutants to the atmosphere via plant transpiration is known as phytovolatilization (Burken and Schnoor 1997). It is a specialized form of phytotransformation that can be used only for those pollutants that are highly volatile in nature. Contaminants like Hg and Se, once they are taken up by plant roots, can be converted into nontoxic forms and volatilized into the atmosphere from the foliage (Limmer and Burken 2016).

1.3.2.2.4   Phytodegradation and the Green Liver Model

Phytodegradation is a process where organic contaminants are taken up by plants and broken down through metabolic processes within the plant tissues or through the effects of enzymes produced by the plants on contaminants in close proximity to the root (Newman and Reynolds 2004). Plants are able to produce various enzymes such as peroxidases, dioxygenases, P450 monooxygenases, laccases, phosphatases, dehalogenases, nitrilases, and nitroreductases that catalyze and accelerate the degradation of complex organic pollutants into simpler molecular forms that are termed *ex planta* metabolic processes. After degradation, simpler molecules are incorporated into plant tissues to aid plant growth. From this point of view, green plants can be regarded as a "green liver" for the biosphere. The "green liver model" has been used to describe phytotransformation, as plants behave analogously to the human liver when dealing with these xenobiotic compounds (Sandermann 1994). In plants, the detoxification of xenobiotics is carried out in three stages: transformation

(Phase-I), conjugation (Phase-II), and sequestration or compartmentation (Phase-III) (Schäffner et al. 2002). In Phase-I, certain initial reactions such as oxidation, reduction, or hydrolysis give rise to xenobiotic compounds that are amenable to subsequent conjugation reactions. In Phase-II, the xenobiotic compound undergoes conjugation with glutathione, sugar, and amino acids. This phase increases the polarity and reduces the toxicity of the xenobiotic compounds. GSH and the glutathione S-transferase (GST) isoenzymes have a crucial role in the degradation of many pesticides, as they are able to form conjugates between various xenobiotics and GSH by nucleophilic addition reactions. In Phase-II, the resulting conjugated compound can undergo various sequestration processes. The sequestration of the xenobiotic occurs within the plant vacuoles or via its binding to the plant cell wall and lignin via ATP-binding cassette (ABC) transporters, whereupon the compound is no longer capable of interfering with cell function. ABC transporters are the best-characterized system to transfer toxic organics out of root cells and into vacuoles after conjugation by GSTs. The xenobiotic can polymerize in a lignin-like manner and develop a complex structure that is sequestered in the plant.

1.3.2.2.5   Rhizofiltration and Rhizoremediation

Rhizofiltration, a root-zone *in situ* or *ex situ* technology, can be used for the elimination of metals from water and aqueous waste streams that are retained only within the roots of aquatic plants. It reduces the mobility of metals and prevents their migration to the groundwater, thus reducing bioavailability for entry into the food chain. Rhizofiltration is particularly effective when low concentrations of contaminants and a large volume of water are involved.

Not only the previously described processes are classified as phytoremediation; some authors also distinguish between indirect and direct phytoremediation. In the case of indirect phytoremediation, plants participate in the detoxification of pollutants via their support of symbiotic, root-associated microbes that actually accomplish contaminant detoxification. This process is called rhizoremediation. In rhizoremediation, plant-supplied substrates stimulate the growth of microbial communities in the rhizosphere to break down organic pollutants in the soil. On the other hand, plants could participate directly through contaminant uptake and subsequent contaminant immobilization or degradation within the plant. Most scientific and commercial interest in phytoremediation now focuses on phytoextraction and phytodegradation where selected plant species are grown on contaminated soils to remove pollutants. At the end of the growth period, plant biomass is harvested, together with the pollutants that have accumulated in their tissues. Depending on the type of contamination, the plants can either be disposed of in a special dump or used in alternative processes, such as burning for energy production (Peuke and Rennenberg 2005). Phytoremediation, in combination with burning the resulting biomass to produce electricity and heat, could become a new eco-friendly form of biotechnology. It is also possible to recover some metals from plant tissue (phytomining), which humans have done for centuries in the case of potassium and which may even become economically valuable (Suman et al. 2018). Phytomining is now a fast-developing field with the potential to generate income by exploiting low-grade ore bodies that are not economical to mine by conventional methods (Nedelkoska and Doran 2000; Peuke and Rennenberg 2005). The overall outcome of a carefully planned phytoremediation-phytomining operation would be a commercially viable metal product (i.e., metal-enriched bio-ore) and land better suited for agricultural operations or general habitation. The success of phytoremediation or phytomining depends on the availability of plant species—ideally those native to the region of interest—able to tolerate and accumulate high concentrations of heavy metals (Peuke and Rennenberg 2005).

Although phytoremediation is a cost-effective, environmentally friendly technology for ecological restoration of contaminated sites, it is generally slower than most other traditional remediation technologies and may take at least several growing seasons to clean up a site. Phytoremediation is also limited by the bioavailability of the pollutant (Marques et al. 2009).

1.3.2.3   Electrokinetic-Phytoremediation

Electrokinetic-phytoremediation technology consists of the application of a low-intensity electric field to the contaminated soil in the close vicinity of growing plants. In this hybrid technology, the removal or degradation of the contaminants is performed by the plant, whereas the electric field enhances the plant activity by increasing the bioavailability of the contaminants by desorption and transport of the contaminants (Cang et al. 2011; Rocha et al. 2019). In addition to the plant's own strategy, an electrokinetic field (EKF) was also recently introduced to enhance the phytoextraction efficiency of metal(loid)-contaminated soils (Aboughalma et al. 2008; Cameselle et al. 2013). Electrokinetic-coupled phytoremediation using *Lemna minor* was tested by Kubiak et al. (2012) to remediate toxic As in water. For this test, artificial As water was prepared using sodium arsenate at a concentration of 150 μg L^{-1}. Their preliminary results showed a higher removal rate of 90% at the end of the experiment. In another study, Pb removal from soil was reported by Hodko et al. (2000), where the electrokinetic (EK) remediation was carried out by applying several electrode configurations to enhance phytoremediation by increasing the depth of soil to prevent the leaching of mobile metals on the ground surface. Bhargavi and Sudha (2015) used an electrokinetic-assisted phytoremediation process to reduce the levels of Cr and Cd. In their study, soil samples were first remediated using the EK method, followed by phytoremediation by extruding the remediated soil samples from the electrokinetic cell. EK-remediated soil was potted to grow the plant *Brassica juncea*. The study reported on 67.43% and 59.78% removal efficiency of Cd and Cr after 25 days of treatment. The EK-remediated soil employed for phytoremediation showed a promising accumulation of Cd and Cr in a single harvest, which was further increased in subsequent harvests. EK-enhanced phytoextraction demonstrated by Mao et al. (2016) removed Pb, As, and Cs from contaminated paddy soil by lowering the pH of the soil to 1.5, which resulted in dissolution of HMs to a large extent, with increased solubility and bioavailability of heavy metals. It was then followed by phytoextraction using plants that enhanced the effectiveness of metal removal from the soil. The optimization of electrical parameters such as electric field intensity, current application mode, distance between the electrodes, and stimulation period and its effect on the mobility and bioavailability of HMs are the associated key challenges (Mao et al. 2016).

1.4   MICROBE-ASSISTED PHYTOREMEDIATION: AN ECO-SUSTAINABLE GREEN TECHNOLOGY FOR ENVIRONMENTAL REMEDIATION

Although phytoremediation is a promising, cost-effective, eco-friendly, and plant-based remediation technology applied to inorganic and organic contaminated soils, water, and sediments all over the world to improve the quality of the environment and human health, it is generally time consuming, with slow degradation, limited uptake, and evapotranspiration of volatile contaminants and phytotoxicity (Peuke and Rennenberg 2005; Ali et al. 2013, 2019). Moreover, plants that are tolerant to the presence of these contaminants often remain relatively small, due to the toxicity of organic and inorganic pollutants that are accumulating or the toxic end product of their degradation. Higher metal concentrations in soil likely inhibit the growth and development of plants and may thus reduce the efficiency of phytoremediation. For successful phytoremediation, a plant should be able to grow rapidly, reach high biomass, and take up large concentrations of heavy metals from the contaminated soils into the roots (Chandra et al. 2018a,b). Plants also require certain and micro- and macronutrients such as nitrogen (N), phosphorus (P), potassium (K), and HMs for their growth and development, and excessive amounts of HMs can become toxic to plants, apart from hyperaccumulators, leading to impaired metabolism, reduced plant growth, and eventually reduced phytoremediation efficacy (Glick 2003, 2012). Recently attention has focused on the role of microbial-assisted phytoremediation in amelioration of plant stress, considered an important tool in the management of contaminated environments (Juwarkar and Singh 2010; Rajkumar et al. 2012). Microbe-assisted phytoremediation utilizes the complex interactions involving plant roots, root exudates, rhizosphere soil, and microbes that result in degradation of contaminants to nontoxic/less-toxic compounds (Finkel et al. 2017). Microbes reduce the toxicity of metal by decomposition or immobilizing the metals from the soil. Microbe-assisted phytoremediation has emerged as a sustainable soil cleanup technology with reduced soil disturbance, low maintenance, and overall low costs (Nanekar et al. 2015; Ali et al. 2017; García-Sánchez et al. 2018). Plants can provide favorable environmental conditions for the colonization and proliferation of microbial communities by offering nutrients through root exudates rich in carbon sources, nutrients, enzymes, and sometimes oxygen (Rajkumar et al. 2012; Redfern and Gunsch 2016; Fatima et al. 2018). Consequently, plant root activities enhance the microbial count and diversity in polluted environments and accelerate the remediation process of pollutants in the rhizosphere as well as the endosphere (Figure 1.3).

The rhizospheric environment is an essential habitat for different microbes, including bacteria, fungi, nematodes, algae, actinomycetes, protozoa, algae, and microarthropods. Such microorganisms exhibit a diversity of associations with plants (Mendes et al. 2013). Microbes have the ability to synthesize and sense signaling molecules that trigger microbial populations to form a biofilm around the root surface and induce a related response (Davey and O'Toole 2000). In many cases, plant-associated microbes are the main contributor to the contaminant

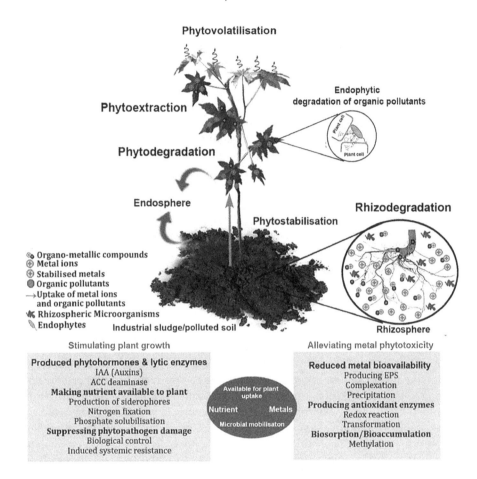

Phytovolatilisation

Phytoextraction

Endophytic
degradation of organic pollutants

Phytodegradation

Plant cell

Plant cell

Endosphere

Rhizodegradation

Phytostabilisation

%₀ Organo-metallic compounds
⊕ Metal ions
⊕ Stabilised metals
◉ Organic pollutants
→Uptake of metal ions
   and organic pollutants
¥ Rhizospheric Microorganisms
❧ Endophytes   Industrial sludge/polluted soil

Rhizosphere

Stimulating plant growth

Produced phytohormones & lytic enzymes
IAA (Auxins)
ACC deaminase
Making nutrient available to plant
Production of siderophores
Nitrogen fixation
Phosphate solubilisation
Suppressing phytopathogen damage
Biological control
Induced systemic resistance

Available for plant uptake

Nutrient   Metals

Microbial mobilisaton

Alleviating metal phytotoxicity

Reduced metal bioavailability
Producing EPS
Complexation
Precipitation
Producing antioxidant enzymes
Redox reaction
Transformation
Biosorption/Bioaccumulation
Methylation

FIGURE 1.3 Various mechanisms of plants and their associated microbes for phytoremediation of heavy metals and organic pollutants from contaminated sites. (Reprinted with permission from Kumar V, Chandra R. 2020b. In: Chandra R, Sobti R (eds) *Microbes for Sustainable Development and Bioremediation*. Boca Raton: CRC Press.)

degradation process and play significant roles in nutrient cycling, improving soil structure, detoxifying harmful contaminants, modulating plant defense responses to stress factors, and assisting in biological control of phytopathogens and plant growth (Braga et al. 2016). These plant growth-promoting activities of root-associated microbes enhance the plant's adaptation and growth in adverse environmental stresses such as drought, salts, HMs, and nutrient deficiency and, ultimately, its phytoremediation efficiency. Consequently, these mutual interactions, also known as the rhizosphere effect, result in elevated numbers, diversity, and metabolic activity of microbes able to degrade contaminants or support plant growth in close vicinity to roots compared to bulk soil (Gkorezis et al. 2016; Ahkami et al. 2017).

1.4.1 PLANT–MICROBE INTERACTIONS: PLANTS AND MICROORGANISMS TOGETHER AT WORK

In the last decade, the ecology of microbes in the rhizosphere was defined by Hiltner (1904) as the active reaction zone of plant roots where physico-chemical and biological processes take place that are induced by the interaction of plants, microorganisms, the soil, and pollutants. Plant roots exert strong effects on the rhizosphere through rhizodeposition (root exudation, production of mucilages, and release of sloughed-off root cells) and by providing suitable ecological niches for microbial growth (Hütsch et al. 2002; Bakker et al. 2013). The rhizosphere is the complex and highly dynamic interface between plant roots and surrounding soil that has a massive amount of soil microorganisms and is the "hot spot." Potential for phytoremediation depends upon the interactions in the rhizosphere among soil, pollutants, microbes, and plants (Segura et al. 2009; Mhlongo et al. 2018). In the rhizosphere, root-based interactions between plants and organisms are highly influenced by edaphic factors. However, the below-ground biological interactions that are driven by root exudates are more complex than those occurring above the soil surface (Olanrewaju et al. 2019). The interface between microbes and plant roots in the rhizosphere is believed to vastly influence the growth and survival of plants. The interactions between plant roots, soil, and microbes significantly change the physico-chemical properties of soil, which in turn alter the microbial population in the rhizosphere. The biotechnological potential of microorganisms to resist and/or remove metals directly from polluted media and their beneficial effects on plant growth may lead to environmentally friendly and cost-effective strategies forreclamation of polluted sites (Mosa et al. 2016).

1.4.2 MICROBES INVOLVED IN ASSISTING PHYTOREMEDIATION

About 20 000 plant species are unable to grow or survive without microbial symbiotic interactions (van der Heijden et al. 2008). This plant–microbe synergism can accelerate the process of remediation of organic and inorganic pollutants but can also increase plant growth and development under adverse environmental conditions (Glick 2003, 2010). Rhizospheric communities also change the plant environment; that is, the microbial degradation of contaminants provides a clean environment by decreasing the pollutant concentration in the area near the roots, favoring plant growth and site restoration (Wei et al. 2017; Guo et al. 2019).

1.4.2.1 Rhizobacteria: Non-Symbiotic Beneficial Rhizospheric Bacteria

In general, plant-associated bacteria migrate from the bulk soil to the rhizosphere of living plants and aggressively colonize the rhizosphere and roots of plants. These so-called rhizobacteria act as an abundant symbiotic partner of plants (Vacheron et al. 2013; Kong and Glick 2017). The term rhizobacteria was coined by Kloepper and Schroth (1978), based on their experiments with radishes (Ahemad and Kibret 2014). They defined rhizobacteria as a community that competitively colonizes plant roots and enhances their growth and also reduces plant diseases. On the basis of their effects on plant growth, rhizobacteria can be classified into beneficial, deleterious, and neutral groups (Glick 2012). The rhizobacteria that stimulate the growth and health of the plant are referred to as plant growth-promoting rhizobacteria (PGPR) (Olanrewaju et al. 2017). It is well established that only 1% to 2% of bacteria promote plant growth in the rhizosphere (Antoun and Kloepper 2001). Beneficial, root-colonizing, rhizosphere bacteria, the PGPR, are defined by three intrinsic characteristics: (i) they must be able to colonize the root; (ii) they must survive and multiply in microhabitats associated with the root surface, in competition with other microbiota, at least for the time needed to express their plant promotion/protection activities; and (iii) they must promote plant growth (Lugtenberg and Kamilova 2009; Kong and Glick 2017). PGPR are known to participate in many important ecosystem processes, such as the biological control of plant pathogens, through various mechanisms, which include production of phytohormones, 1-aminocyclopropane-1-carboxylate (ACC) deaminase, osmoprotectants, siderophores, organic acids, nutrient cycling, and/or seedling growth (Beneduzi et al. 2012; Vacheron et al. 2013; Ahemad and Kibret 2014; Gouda et al. 2018).

1.4.2.2   The Mutualistic Symbionts: N_2-Fixing Bacteria

Dinotrogen (N_2)-fixation is the first step for cycling N to the biosphere from the atmosphere, a key input of N for plant productivity. Symbiotic N_2-fixation is a well-known process exclusively driven by bacteria, the only organisms possessing the key enzyme nitrogenase, which specifically reduces atmospheric N_2 to ammonia in symbiotic root nodules. The bacteria belong to the genera *Rhizobium, Azorhizobium, Bradyrhizobium, Sinorhizobium,* and *Mesorhizobium* and are collectively termed rhizobia (Gage 2004). These bacteria interact with legume roots, leading to the formation of N_2-fixing nodules. Other bacteria (actinomycetes) of the genus *Frankia* form nodules on the root of "actinorhizal" plant species, which are of great ecological importance.

1.4.2.3   Endophytic Bacteria

Endophytic bacteria can be defined as bacteria that colonize the internal tissue of the plant without causing visible external signs of infection or a negative effect on their host plants or environment (Doty 2008; Weyens et al. 2009c,d). In general, endophytic bacteria occur at lower population densities than rhizospheric bacteria or bacterial pathogens. They have established harmonious associations with host plants during symbiotic, mutualistic, commensalist, and trophobiotic relationships over a long evolutionary process (Lodewyckx et al. 2002). It is well established that the soil environment is the main source of endophytes. Endophytic bacteria in a single plant host are not restricted to a single species but comprise several genera and species. Because endophytic bacteria can proliferate within the plant tissue, they are likely to interact closely with their host and therefore face less competition for nutrients and be more protected from adverse changes in the environment than rhizospheric and phyllosphere bacteria (Ashraf et al. 2018a,b). Many endophytic bacteria, particularly those inhabiting plants growing in a polluted environment, produce degradative enzymes and contribute to the degradation of several types of organic compounds present in the rhizosphere and endosphere (Weyens et al. 2009a,b,c,d, 2010a,b; Khan et al. 2014a,b; Ma et al. 2016a,b; Hussain et al. 2018).

1.4.2.4 Fungi: Non-Symbiotic Beneficial Rhizospheric Fungi

Fungi have been defined as eukaryotic, heterotrophic, and absorptive organisms, which typically develop a branched, tubular-like structure called a mycelium and reproduce by means of sporulation (Jambon et al. 2018). They play an important role in element cycling, rock and mineral transformations, bio-weathering, mycogenic mineral formation, and organic and inorganic transformation (Hata et al. 2010; Hodge 2014). The extensive hyphal networks in the soil can also significantly contribute to the stabilization of soil aggregates; they modify the chemical composition of root exudates and soil pH and control metal bioavailability in the soil. Fungi can efficiently explore soil microsites that are not accessible for plant roots due to the small diameter of the mycelia and can compete with other microorganisms for water and metal uptake, protect roots from direct interaction with the metals, and hinder metal transport through increased soil hydrophobicity (Leyval et al. 2002; Miransari 2011; Yang et al. 2016). Many fungi, such as *Trichoderma*, *Aspergillus*, and the arbuscular mycorrhizal fungi (AMF), have shown the potential to improve phytoremediation processes in metal-contaminated soils because they have a high ability to immobilize toxic HMs (Rajtor and Piotrowska-Seget 2016; Schneider et al. 2016; Yang et al. 2017). AMF are ubiquitous, obligate microbial symbionts, unable to complete their life cycle without colonizing a host plant, whose origin and divergence have been dated back to more than 450 million years ago. AMF are thought to be one of the most important soil microbial groups that affect metal uptake by plants and metal immobilization in soils and are commonly introduced into the soil for land reclamation (Fecih and Baoune 2019). In some cases, AMF contributed to significant storage of metal(loid)s at the root level, instead of the above-ground tissues of the host plants (Gaur and Adholeya 2004; Meier et al. 2012). However, in other cases, AM fungi contributed to enhanced uptake and translocation to shoots, promoting phytoextraction success (Cabral et al. 2015; Elhindi et al. 2018).

1.4.3 GENERAL MECHANISM OF MICROBE-ASSISTED PHYTOREMEDIATION

Mechanisms of plant-growth promotion by plant-associated microbes vary greatly and can be broadly categorized into direct and indirect effects, as discussed subsequently.

1.4.3.1 Direct Promotion of Phytoremediation

Direct plant-growth promotion is based on either the stipulation of the plants with favorable microbial compounds or improving the nutrient uptake by the plant from the environment. Rhizospheric or endophytic microbes promote plant growth and development through a variety of mechanisms, including the fixation of atmospheric N_2 and supply of it to plants; synthesis of iron chelators referred to as siderophores; synthesis and release of different phytohormones, including auxins, cytokinins, and gibberellins, which can enhance various stages of plant growth as well as favoring the establishment of rhizobial or mycorrhizal symbioses; improved ammonia production; and solubilization of inorganic phosphate and mineralization of organic phosphate and/or other nutrients, which then become more readily available for plant growth (Glick 2010, 2012; Huo et al. 2012; Ma et al. 2016a,b). In addition to this, plant growth-promoting (PGP) microbes enhance the tolerance capacity of the plant to a variety of environmental stresses through the production of ACC deaminase that can modulate plant growth and development (Glick 2014).

1.4.3.2 Indirect Promotion of Phytoremediation

Indirect promotion of plant growth occurs when PGP microbes prevent the deleterious effects of phytopathogens, usually fungi and nematodes, through the production of antimicrobial compounds, thereby controlling diseases and competition for iron and nutrients or colonization sites, to mention a few mechanisms (Glick 2003, 2010, 2012). Pathogen suppression may be achieved through a variety of mechanisms like production and release of cyanide, antibiotics, or extracellular lytic enzymes including chitinases, proteases, β-1,3 glucanases, cellulases, and laminarinases or competition for nutrients and niches in the rhizosphere, parasitism, and predation. Moreover, some PGP microbes secrete antibiotics, which are particularly relevant for rhizosphere and rhizoplane colonization (van Loon and Bakker 2005; Glick 2012). Many PGP microbes act antagonistically toward plant pathogens by producing antimicrobial compounds or by interfering with virulence factors via effectors delivered by type-3-secretion systems.

2 Plant–Microbe Partnership

Multipurpose Benefits and Role in Assisting Phytoremediation

2.1 INTRODUCTION

Soil, water, and air pollution is a major concern of the entire world that has increased in recent years and can cause damage to human health, soil productivity, and ecosystem and ecological health. Phytoremediation is a promising, environment-friendly technique to remove pollutants from the site contaminated with toxic pollutants. This green remediating technology is based on the combined action of plants and their associated microbial communities to degrade, remove, transform, or immobilize toxic compounds located in soils, sediments, groundwater, and surface water (Garbisu et al. 2002; Padmavathiamma and Li 2007). In phytoremediation processes, plants and microbes coexist or compete for survival and growth, and their synergic interactions play a vital role in adapting to contaminated environments and can thus be explored to improve microbe-assisted phytoremediation (Gerhardt et al. 2015; Hrynkiewicz et al. 2018). Microbes can facilitate phytoremediation in different manners: expedite plant biomass; increase or decrease pollutant availability in soil, sludge, sediment, or wastewater; and enhance metal translocation from water, soil, sludge, or sediment to roots or from root to shoot tissues (Karthikeyan and Kulakow 2003; Dotaniya et al. 2018). This chapter attempts to review the recent advances and applications in understanding the biochemical and molecular mechanisms of plant–microbe interactions and their role in phytoremediation processes. This chapter also addresses some aspects of the taxonomical and functional microbial ecology of the rhizosphere. Further, we will elaborate on the mechanisms underlying plant–microbe–metal interactions in the rhizosphere, namely: (i) plant–microbe interactions (molecular signaling, quorum sensing [QS], and establishment of associative symbiosis) and (ii) heavy metal (HM) versus plant–microbe interactions (role of plant–microbe–metal interactions in HM detoxification, mobilization, immobilization, transformation, transport, and distribution).

2.2   THE RHIZOSPHERE AND ITS ECOLOGY

The rhizosphere is known to be a hot spot for numerous microbial activities and is considered one of the most complex and dynamic ecosystems on Earth (Raaijmakers et al. 2009). The term "rhizosphere" is derived from the Greek word "rhiza," meaning root, and "sphere," meaning field of influence. Lorenz Hiltner (1904) was the first scientist to mention the concept of the rhizosphere as a zone surrounding the roots of plants in which complex relationships exist among the plant, the soil microbes, and the soil itself. The rhizosphere has been broadly subdivided into the three following zones (McNear Jr 2013; Prashar et al. 2014): (i) the endorhizosphere: part of the root tissue (endodermis and cortical layers); (ii) the rhizoplane: the root surface where soil particles and microbes adhere, which has an epidermis layer, a middle layer cortex, and a polysaccharide layer; and (iii) the ectorhizosphere: the zone where the roots are adjacent to the soil surface. Apart from these three basic zones, certain other layers may be defined in some cases; for example, in plants with mycorrhizal associations, there is a zone termed the mycorrhizosphere (Johansson et al. 2004). The interface between microbes and the rhizosphere is considered to greatly influence the growth and survival of plants at contaminated sites, and understanding its ecology and evolution is key to enhancing plant productivity and ecosystem functioning (Philippot et al. 2013; Ahkami et al. 2017). The soil in the rhizosphere supports a typically diverse and densely populated microbial community including bacteria, fungi, oomycetes, nematodes, protists, algae, viruses, and archaea (Mendes et al. 2013). The corresponding microbial community associated with plant roots can be referred to as the rhizomicrobiome. The composition of a microbial community in a rhizosphere is also known to differ both qualitatively and quantitatively from that in non-rhizospheric soil (Prashar et al. 2014). The rhizospheric microbial load ranges from 10^{10} to 10^{12} per gram of soil, while it is generally less than 10^8 in bulk soil (Foster 1988). The actual composition of a rhizospheric microbial community depends on root type, plant species, plant age, and soil type, as well as other factors such as exposure history of the plant roots to pollutants (Lareen et al. 2016; Qiao et al. 2017; Wang et al. 2017). Microbes present in the rhizosphere play important roles in plant growth and ecological fitness of their plant host. Rhizosphere microbes that have been well studied for their beneficial effects on plant growth and health are nitrogen-fixing bacteria and archaea, arbuscular mycorrhizal fungi (AMF), ectomycorrhizal fungi, plant growth-promoting rhizobacteria (PGPR), biocontrol microorganisms, mycoparasitic fungi, and protozoa (Mendes et al. 2013; Lareen et al. 2016). Rhizosphere organisms that are deleterious to plant growth and health include pathogenic bacteria, fungi, oomycetes, nematodes, and microarthropods (Mercado-Blanco et al. 2018). The third group of microorganisms that can be found in the rhizosphere is human pathogens; in particular, members of the family *Enterobacteriaceae* can invade root tissue. Important microbial processes that are believed to occur in the rhizosphere include pathogenesis and its counterpart, plant protection, as well as the production of antibiotics, geochemical cycling of minerals, and plant colonization (Garcia and Kao-Kniffin 2018). Thus, the interaction of plants and rhizobial microbes is not only key to global biogeochemical cycles but also essential for the removal of harmful contaminants from polluted environments (Segura et al. 2009; Abhilash et al. 2012).

Rhizosphere interactions are affected by many different regulatory signals, of which only a few have been identified, recalling a quote by Leonardo da Vinci: "We know better the mechanics of celestial bodies than the functioning of the soil below our feet." Rhizosphere interactions are very dynamic and can be altered by addition or loss of any of the players. Root exudates and microorganisms are important components of rhizosphere ecology and play important roles in changing the bioavailability of metals and nutrients (Kim et al. 2010; Luo et al. 2014; Montiel-Rozas et al. 2016).

2.2.1 BACTERIA

Bacteria (singular: bacterium) are omnipresent and successfully occupy ecological niches as well as colonization hotspots like the plant rhizosphere. The concentration of bacteria in the rhizosphere is 10 to 1000 times higher than that in bulk soil, but it is still 100-fold lower than that in the average laboratory medium (Lugtenberg and Kamilova 2009). It is reported that up to 15% of the total root surface may be covered by a variety of bacterial strains (van Loon 2007). The most common genera of bacteria that have been found in the rhizosphere are *Pseudomonas, Bacillus, Arthrobacter, Rhizobia, Agrobacterium, Alcaligenes, Azotobacter, Mycobacterium, Flavobacterium, Cellulomonas*, and *Micrococcus* (Ahemad and Kibret 2014). The discovery of HM-resistant and organic compounds degrading rhizospheric bacteria that are able to promote plant growth in highly polluted environments has raised hopes for reclamation of polluted soil (Lugtenberg and Kamilova 2009; Gouda et al. 2018). Plant growth-promoting effects by rhizobacteria can greatly improve plant performance and also result in higher amounts of accumulated trace elements (Kong and Glick 2017). The community structure of bacteria in the rhizosphere is influenced by a variety of biotic and abiotic factors. The plant itself is the most crucial factor in determining the predominant bacterial strains in the rhizosphere due to the significant role of the root exudates in setup of bacterial populations. Also, plant-related features such as the cultivars, age, and development of plant and root characteristics play a critical role in deciding the rhizosphere community structure of bacteria (Glick 2012).

2.2.2 FUNGI

A fungus (plural: fungi) is any member of the group of eukaryotic organisms that includes microorganisms such as yeasts and molds, as well as the more familiar mushrooms. Rhizosphere fungi are closely linked to plant health and growth owing to their roles in antagonizing pathogens, decomposing plant residues, and providing nutrients (Jones et al. 1991; Cabral et al. 2015; Jambon et al. 2018). The structure of fungal communities in the rhizosphere is the result of complex interactions among selection factors that may favor beneficial or detrimental relationships. Concentration gradients in oxygen and organic exudates attract fungi toward the rhizosphere, where they can feed and grow (Meier et al. 2012; Begum et al. 2019). Mycorrhizal fungi, which belong to the phylum Glomeromycota, form mutualistic symbiotic associations with roots; ectomycorrhizal fungi (some of whose above-ground fruiting bodies we call mushrooms) form a mat outside the root and penetrate only the intercellular spaces; and AMF are endomycorrhizal; that is, they live inside the living cells of the root cortex (Leyval et al. 2002; Miransari 2011). AMF are ubiquitous soil microflora and constitute an important functional component of the rhizosphere. These fungi form symbiotic relationships with the roots of 80%–90% of vascular plant species in natural, agricultural, and forest ecosystems and account for 5% to 50% of the total biomass of soil microbes. Such associations are also common in aquatic plants under oligotrophic conditions. Arbuscular mycorrhizal symbiosis is 400–446 million years old and the most widespread type of mycorrhizal association with plants possessing true roots, that is, pteridophytes, gymnosperms, and angiosperms. It is estimated that AMF obtains 4%–20% of photosynthetically fixed total carbon and in exchange provides plants with nitrogen (N), phosphorus (P), and other nutrients, which are usually less available in soil polluted with heavy metals and hydrocarbons. The prospect of fungal symbionts existing in metal-contaminated soils has important implications for phytoremediation (mycorrhizoremediation) of metal-contaminated soils, as AMF helps plant growth through enhanced nutrient uptake (Khan 2006).

2.2.3 ARCHAEA

Archaea are "newcomers" in biology and form the third domain of life (Woese and Fox 1977). Under specific conditions, such as reduced oxygen and/or high CO or CO_2 pressure, they may become very abundant and active contributors to rhizosphere processes. While few of their roles are now known, archaea may largely be unsung heroes, participating in nutrient turnover and sustaining important ecological functions in plant roots. Archaea represent a significant component of the plant microbiome, but their function is still unclear. The factors influencing archaeal functionality under specific anaerobic conditions in rice roots have been analyzed, but their ecological roles and interactions with plants have remained largely unclear. The fact that most archaea are difficult to cultivate and that plant-associated archaeal pathogens are currently not known may be attributed to the lack of knowledge. Within the rhizosphere, methane-producing archaea (methanogens) in rice paddies, wetlands, lake and ocean sediments, and other anoxic and flooded sediments are involved with the production of the highly potent greenhouse gas methane.

2.2.4 PROTOZOA

Protozoas (also called protozoan) are single-celled animals that feed primarily on bacteria but also eat other protozoans, soluble organic matter, and sometimes fungi. They can affect plant health by mineralizing nutrients and altering the structure and activity of root-associated communities. In this context, bacteria in the rhizosphere are strongly top-down-regulated by the grazing abilities of protozoans. For instance, the grazing abilities of protozoans may promote the production of plant growth hormones or enhance the survival of beneficial microbes, suppressing pathogens. The most important bacterial grazers in the soil are naked amoebae due to their high biomass and turnover and specialized feeding modes. Amoebae are grazing bacterial biofilms and colonies attached to soil and root surfaces and thus have access to the majority of bacteria in soil. Protozoans are also an important food source for other soil organisms and help to suppress disease by competing with or feeding on pathogens. They are widespread in wetland rhizospheres, where their feeding on roots and microorganisms, on the one hand, enhances nutrient cycling through the microbial loop and, on the other hand, facilitates the passage of nutrients and energy up the food chain to larger organisms and to higher trophic levels; furthermore, their movement physically transports nutrients and bacteria within the rhizosphere.

2.3 INTERACTIONS IN THE RHIZOSPHERE

Prokaryotes and eukaryotes have coexisted for millions of years on Earth. It is estimated that humans have 10^{13} human cells and 10^{14} bacterial cells, including the endogenous bacterial flora. As a result of this long association, prokaryotes have developed both beneficial and detrimental relationships with eukaryotes. The plant root system, which was traditionally thought to provide anchorage and uptake of nutrients and water, is a chemical factory that mediates numerous underground interactions (Nie et al. 2011; Ma et al. 2016a; Hassani et al. 2018). These include mutualistic associations with beneficial microbes, such as rhizobia, mycorrhizae, endophytes, and PGPR, and parasitic interactions with other plants, pathogenic microbes, and invertebrates (Weyens et al. 2009a,b; Mendes et al. 2013; Kong and Glick 2017; Ashraf et al. 2018a,b). Interactions between plants, rhizospheric microbes, and soils have shaped terrestrial ecosystems during the Earth's evolution, and specific communication and recognition processes have been established between plants and microbes through co-evolution (Yang et al. 2016; Mhlongo et al. 2018). These specific recognition processes have been well studied in pathogens and in some PGPR, but little is known about pollutant-degrading rhizobacteria.

2.3.1 PLANT–MICROBE INTERACTIONS

As a result of microbial colonization in and around growing plant roots, various kinds of relationships, such as associative, symbiotic, neutralistic, or parasitic, may develop, depending upon factors like nutrient status of the soil, overall soil environment, plant defense mechanisms, and certainly the proliferating microorganism itself (Tarkka et al. 2013; Braga et al. 2016; Yang et al. 2016; Jambon et al. 2018; Mhlongo et al. 2018). Beneficial plant–microbe interactions are symbiotic interactions in which costs and benefits are shared by the plants and microorganisms. This beneficial interaction can be divided into two categories: (i) mutualistic interactions and (ii) cooperation (also known as associative symbioses). Mutualistic interactions correspond to intimate and mostly obligate interactions between microbes and a restricted range of compatible host plants. This interaction generally leads to the formation of a specific structure, for example, nodules during the symbiosis between nodulating rhizobia and Fabaceae and arbuscules in endomycorrhizal symbiosis. On the other hand, associative symbioses correspond to less obligate and specific interactions where soil microbes are able to colonize the surface of the root system (and sometimes root inner tissues) and to stimulate the growth and health of the plant and are referred to as plant growth-promoting microbes (PGPMs) (van Loon 2007; Zaidi et al. 2009; Bashan et al. 2013; Ahemad and Kibret 2014; Ahemad 2019). Colonization of plant host roots by PGPMs is heterogeneous along the root system; their competitiveness regarding this process is a sine qua non for plant-growth promotion. Soil conditions play an important part in the type of interaction that soil will have with microbes. The soil environment consists of a variety of physical, biological, and chemical factors that affect the abundance, diversity, and interaction of a microorganism with plant root rot (Ma et al. 2016a,b). However, the specificity of the plant–microbe interaction is dependent upon soil conditions, including organic matter, pH, temperature, nutrients, and pollutant level, which can alter contaminant bioavailability, root exudate composition, and nutrient levels (Karthikeyan and Kulakow 2003).

2.3.1.1   Root Exudates

Plant and microbes can interact with one another in a variety of different ways to absorb, degrade, or remove toxic contaminants from polluted environments. Plant–microbes communication is mediated by the root exudates through the chemotactic response of the microorganism towards exudates like sugars, flavanoids, organic acids, and amino acids, leading to root colonization (Lareen et al. 2016; Montiel-Rozas et al. 2016). The ability of microbes to "sense" plant-derived compounds present in root exudates or to communicate with other microbes plays an important role—not only in colonization and competition with other microbes but also in expression of catabolic genes. Root exudates are the largest source of carbon supply within soil and the most import part of rhizodeposits, and they play an active and relatively well-documented role in the regulation of symbiotic and protective interactions for a tremendous diversity of microorganisms (Luo et al. 2014; Olanrewaju et al. 2019). It is thought that plants release up to 40% of their photosynthetic fixed carbon through the roots into the surrounding area due to this so-called rhizodeposition. Generally, root exudates are transported across the plasma membrane and secreted into the surrounding rhizosphere by plant roots via either passive or active mechanisms. Root exudates have been grouped into low molecular weight (LMW) compounds and high molecular weight (HMW) compounds. LMW compounds, that is, amino acids, organic acids, sugars, phenolics, and various other secondary metabolites, are believed to make up the majority of root exudates, whereas HMW exudates primarily include mucilage and proteins (Dennis et al. 2010). Table 2.1 summarizes the main root exudates released by different plants.

2.3.1.2   Quorum Sensing and Rhizospheric Chemical Dialogue

The rhizosphere is a crossroads for nutrient exchange between plants and soil microbes, and this exchange is likely mediated by bacterial signaling, or QS. QS is a density-dependent regulatory mechanism used by many bacteria to regulate gene expression in a coordinated manner. It was first described in the Gram-positive aquatic bacterium *Vibrio fischeri* as the signal-mediated induction of the *lux* genes responsible for bioluminescence. Its activation is mediated by small autoinducer molecules, N-acyl homoserine lactones (AHLs), which are responsible for cell-cell communication and the coordinated action of many bacteria, including plant-associated bacteria. Plant roots discharge plenty of secondary metabolites such as terpenes, flavonoids, glucosinolates, and phenylpropanoids into the rhizosphere. Plant addition of resources to the rhizosphere permits increased cell density and activity, which are reflected in the increased abundance of the QS signal AHLs, regulating QS and QS-controlled

TABLE 2.1
Organic Compounds Identified in Root Exudates of Different Plants

Components	Identified Compounds
Sugars	Glucose, fructose, galactose, sucrose, arabinose, maltose, mannose, mucilages of various compositions, oligosaccharides, raffinose, rhamnose, ribose, xylose, deoxyribose
Amino acids	α-alanine, β-alanine, g-aminobutyric acid, α-aminoadipic acid, arginine, asparagine, aspartic acid, citrulline, cystathionine, cysteine, cystine, deoxymugineic acid, 3-epi-hydroxy-mugineic acid, glutamine, glutamic acid, glycine, histidine, homoserine, isoleucine, leucine, lysine, methionine, mugineic acid, ornithine, phenylalanine, proline, serine, threonine, tryptophan, tyrosine, valine
Organic acids	Acetic, aconitic, ascorbic, aldonic, benzoic, butyric, caffeic, citric, erythronic, ferulic, formic, fumaric, glutaric, glycolic, lactic, glyoxylic, malic, malonic, oxaloacetic, oxalic, p-coumaric, p-hydroxybenzoic, piscidic, propionic, pyruvic, succinic, syringic, tartaric, tetronic, valeric, and vanillic acids
Fatty acids	Linoleic acid, linolenic acid, oleic acid, palmitic acid, stearic acid
Sterol	Campesterol, cholesterol, sitosterol, stigmasterol
Growth factors and vitamins	p-Aminobenzoic acid, biotin, choline, N-methyl nicotinic acid, niacin, pantothenic acid, thiamine, riboflavin, pyridoxine, pantothenate
Enzymes	Amylase, invertase, peroxidase, phenolase, acid/alkaline phosphatase, polygalacturonase, protease
Flavanones and purine nucleotides	Adenine, flavanone, guanine, uridine/cytidine
Miscellaneous	Auxins, scopoletin, hydrocyanic acid, glucosides, unidentified ninhydrin-positive compounds, unidentifiable soluble proteins, reducing compounds, ethanol, glycinebetaine, inositol and myo-inositol-like compounds, Al-induced polypeptides, dihydroquinone, sorgoleone, isothiocyanates, inorganic ions and gaseous molecules (e.g., CO_2, H_2, H, OH, HCO_3), some alcohols, fatty acids, alkyl sulfides

behaviors. Flavonoids, for instance, are able to mimic QS molecules and thereby influence the bacterial metabolism. AHLs-mediated QS is likely to be important in the rhizosphere due to the apparent dominance of Proteobacteria among rhizosphere microbial communities. Thus, the QS system controls basic processes of bacterial life, for example, biofilm formation and motility, and is likely also to affect the quality and quantity of volatiles. This is particularly important in highly competitive situations between different bacterial organisms that benefit from nutrient-rich conditions in the rhizosphere. Bacterial species have adapted to these ever-changing conditions and are capable of starting colonization by forming microcolonies on different parts of the roots from the tip to elongation zone. Such microcolonies eventually grow into large population sizes on roots to form mature biofilms. Root exudates serve as a major plant-derived factor responsible for triggering root colonization and biofilm associations. The plant growth-promoting pseudomonads have been reported to discontinuously colonize the root surface, developing as small biofilms along epidermal fissures. Production of extracellular polymeric substances (EPSs) is generally important in biofilm formation and likewise can affect root colonization in many bacterial species, such as for *Azospirillum brasilense, Gluconacetobacter diazotrophicus, Herbaspirillum seropedicae, Agrobacterium tumefaciens, Sinorhizobium meliloti*, and *Pseudomonas fluorescens. Bacillus subtilis*, a plant growth-promoting Gram-positive bacterium widely used as a biofertilizer, exhibits differential biofilm formations on *Arabidopsis thaliana* root surfaces.

2.4 PLANT–MICROBE PARTNERSHIP TO IMPROVE PHYTOREMEDIATION

2.4.1 HEAVY METAL MOBILIZATION IN THE RHIZOSPHERE

Plants growing in HM-contaminated sites harbor a diverse group of microorganisms that are capable of tolerating a high concentration of HMs and providing a number of benefits to both the soil and plants (Gaur and Adholeya 2004; Juhanson et al. 2007; Hou et al. 2015; Ren et al. 2019). The availability of HMs for plant metabolisms is related to the chemical forms of metals, soil–plant and water–plant interaction, and microbial activities because a large proportion of HMs are bound to various organic and inorganic constituents of polluted soil and their phytoavailability is closely related to their chemical speciation (Wu et al. 2006b; Rajkumar et al. 2012; Thijs et al. 2017). The soil microorganisms play a significant role in affecting HM mobility and availability to the plant through the release of chelating agents, acidification, phosphate solubilization, and oxidation/redox change mechanisms (Ma et al. 2016a; Ojuederie and Babalola 2017). The various processes, such as chemical transformation, chelation, and protonation, will lead to mobilization of HMs, whereas precipitation or sorption decreases metal availability. Among the various rhizospheric microbes, rhizospheric bacteria deserve special attention because they can directly improve the phytoremediation process by altering the solubility, availability, and transport of HMs through altering the soil pH, release of chelators and oxidation/ reduction reactions in the rhizosphere (Lugtenberg and Kamilova 2009; Liu et al. 2013; Kong and Glick 2017; Gouda et al. 2018).

2.4.1.1 Siderophores

Most plant and plant-associated microbes (i.e., bacteria and fungi) can produce LMW (200–2000 Da) iron (Fe) chelators called siderophores in response to Fe-limiting conditions. In addition to Fe, siderophores can also form stable complexes with other metals such as zinc (Zn), silver (Ag), copper (Cu), aluminium (Al), manganese (Mn), arsenic (As), cadmium (Cd), chromium (Cr), cobalt (Co), magnesium (Mg), mercury (Hg), lead (Pb), and radionuclides, including plutonium (Pu) and uranium (U), with variable affinities. Supply of Fe to growing plants under HM stress becomes more important, and bacterial siderophores help to minimize the stress imposed by HM contamination (Rajkumar et al. 2010; Mosa et al. 2016). Currently, ~500 different siderophores have been identified by various groups of researchers (Rajkumar et al. 2010). Although siderophores differ widely in their overall structure, the functional groups that coordinate the Fe atom are not as diverse. Interestingly, the binding affinity of phytosiderophores for Fe is less than the affinity of microbial siderophores, but plants require a lower Fe concentration for normal growth than do microbes. Siderophores play a significant role in metal mobilization and accumulation in plants and enhance the phytoremediation efficacy of plants (Rajkumar et al. 2010; Ahmed and Holmström 2014).

2.4.1.2 Organic Acids

Numerous LMW organic acids (i.e., gluconic acid, oxalic acid, tartaric acid, citric acid, acetic acid, 2-ketogluconic acid, and 5-ketogluconic acid) produced by plant-associated microbes play an important role in the complexation of toxic HMs and essential mineral nutrients and increase their mobility in the rhizosphere for plant-root uptake (Rajkumar et al. 2010). Organic acids are CHO-containing compounds characterized by the presence of one or more carboxyl groups with a maximum molecular weight of 300 daltons. In general, organic acids can bind metal ions in soil solution by complexation reactions. However, the stability of the ligand:metal complexes is dependent on several factors such as the nature of organic acids (number of carboxylic groups and their position), the binding form of the HMs present, and the pH of the soil solution. Saravanan et al. (2007) demonstrated the potential of *Gluconacetobacter diazotrophicus* to produce a gluconic acid derivative, that is, 5-ketogluconic acid, which aids in the solubilization of Zn compounds. Experiments with rhizosphere bacteria of a Cd/Zn-hyperaccumulating plant, *Sedum alfredii*, also revealed that the inoculation of soil with Cd/Zn-resistant rhizosphere bacteria significantly increased the water-soluble Zn and Cd concentrations when compared with uninoculated controls. In this case, enhanced HM mobilization could be correlated with the increased production of formic acid, acetic acid, tartaric acid, succinic acid, and oxalic acid.

2.4.1.3 Biosurfactants

Biosurfactants are amphiphilic compounds produced in living spaces or excreted extracellular hydrophobic and hydrophilic moieties that confer on organisms the ability to accumulate between fluid phases, thus reducing surface and interfacial tension. Some plant-associated microbes include *Acinetobacter* sp., *Bacillus* sp., and *Candida antartica*; can produce biosurfactants that form complexes with heavy metals at the soil interface, desorbing metals from the soil matrix, and thus enhance the solubility and bioavailability of HMs; and may be useful for its phytoremediation (Rajkumar et al. 2012). For instance, Juwarkar et al. (2007) verified metal mobilization through biosurfactants produced by *P. aeruginosa* BS2, which aided in the solubilization of Pb and Cd.

2.4.1.4 Polymeric Substances and Glycoproteins

The production of EPSs, mucopolysaccarides, and proteins by plant-associated microbes can also play a significant role in complexing toxic HMs and decreasing their mobility in soil (Gupta and Diwan 2017). EPSs produced by rhizobacteria mainly consist of polysaccharides, proteins, humic substances, uronic acid, nucleic acid, lipids, and glycoproteins surrounding the cells; depict a strong binding capacity to heavy metals; and entrap precipitated metal sulfides and oxides, leading to the development of EPS-metal complexes and subsequently enhancing HM remediation (Pal and Paul 2008; Joshi and Juwarkar 2009). Joshi and Juwarkar (2009), for instance, investigated the immobilization of Cd and Cr after inoculation of EPSs-producing *Azotobacter* spp. and found that these isolates were able to bind 15.2 mg g^{-1} of Cd and 21.9 mg g^{-1} of Cr. Moreover, they also observed that the addition of *Azotobacter* to HM-contaminated soils decreased Cd (-0.5) and Cr (-0.4) uptake by *Triticum aestivum*. Similarly, Gonzalez-Chavez et al. (2004) assessed the ability of AMF-produced insoluble glycoprotein and glomalin to form complexes with HMs and found that up to 4.3 mg g^{-1} Cu, 1.1 mg g^{-1} Pb, and 0.1 mg g^{-1} Cd of glomalin could be extracted from HMs-contaminated soils. Since there is a correlation between the amount of glomalin in the soil and the amount of HMs bound, AMF strains with significant secretion of glomalin should be more suitable for phytostabilization efforts.

2.4.2 MECHANISM OF HEAVY METAL TOLERANCE IN MICROBES

A higher amount of metals, whether essential or nonessential, in the environment is harmful to microbes, plants, animals, and humans. At higher concentrations, metals can damage the membranes of microbial cell walls and interrupt the function of cells by damaging DNA structure and altering enzyme specificity. However, some HM-resistant microbes are adaptive to HM-rich environments. For survival under a metal-stressed environment, microbes have evolved several mechanisms by which they can immobilize, mobilize, or transform metals, rendering them inactive to tolerate the uptake of heavy metal ions and help them function well enough in contaminated environments (Khan et al. 2009; Gupta and Diwan 2017; Ahemad 2019). The possible mechanisms of metal resistance systems in microbes are identified and include elimination through a permeability barrier, capturing and sequestering in the cell (either intra- or extracellular means), enzymatic reduction, active efflux pumps, and diminution in sensitive cellular targets to metal. The energy-dependent active efflux of toxic metal ions (this process is known as extrusion) in which metals are pushed out of the cell through chromosomal/plasmid-mediated events is mostly recognized in the largest group of metal-resistant microbes.

A large number of bacteria are known to possess efflux transporters that excrete toxic metals that are present in a too-high concentration out of the cell via ATPase pumps or chemiosmotic ion/proton pumps. These types of transporters are characterized by a high substrate affinity, and they are therefore able to reduce the metal burden in the cytosol. For instance, plasmid-encoded and energy-dependent metal efflux systems involving ATPases and chemiosmotic ion/proton pumps are also reported for As, Cr, and Cd resistance in other bacteria (Roane and Pepper 2000).

Binding of metals to extracellular material can immobilize the metal and prevent its entry into the cell. For example, metal binding to anionic functional groups (e.g., sulfhydryl, carboxyl, hydroxyl, sulfonate, amine, and amide groups) present on microbial cell surfaces occurs for most metals and prevents their intake into a bacterial cell. Similarly, microbial extracellular polymers, such as polysaccharides, proteins, and humic substances, can efficiently bind HMs. These substances thus detoxify metals simply by complex formation or by forming an effective barrier surrounding the cell and protecting the microbial cell against toxic substances. For instance, EPSs, a complex blend of HMW biopolymeric metabolite secreted by rhizobacteria, fungi, a few plants, and microalgae, not only protect cells against dewatering or toxic substances but serve as a carbon and energy source, too (Sessitsch et al. 2013; Gupta and Diwan 2017).

The interaction between EPSs and HMs appears very complex in the form of electrostatic attraction in which surface complex formation and chemical interaction between HM ions and the functional groups of EPSs occur (Dobrowolski et al. 2017). It has been reported that due to the presence of the acyl group, EPSs show anionic properties, which increase the interaction with other cationic HMs (Cd^{2+}, Pb^{2+}, Co^{2+}, and Ni^{2+}) and form EPS-metal complexes. These substances thus detoxify metals simply by complex formation or by forming an effective barrier surrounding the cell (Rajkumar et al. 2010). Also, siderophores secreted by rhizosphere bacteria and fungi have an important role in the acquisition of several heavy metals. Siderophores can also diminish metal bioavailability and, in turn, its toxicity by binding metal ions that have chemistry akin to that of iron (Dimkpa et al. 2008; Rajkumar et al. 2010). Sometimes, many bacteria mediate reactions or produce metabolites that result in crystallization and precipitation of HMs.

In addition, several bacteria have developed a cytosolic sequestration mechanism for defense from HM toxicity. This mechanism can constitute an effective detoxification process, and the respective microbes might be able to accumulate metals in high intracellular concentrations. Once inside the cell, metal ions might also become compartmentalized or converted into more innocuous forms. This process of the detoxification mechanism in microbes facilitates metal accumulation in high concentrations (Ahemad 2019). For this, a marvelous example is a synthesis of metallothioneins, which are low molecular weight polypeptides that bind and sequester multiple metal ions, such as Cd, Zn, Hg, Cu, and Ag, and are found throughout animals and plants as well as in some prokaryotes. The production of these novel metal-detoxifying proteins is induced in the presence of metal stress. In order to enhance, sequester, or accumulate HMs, bacteria with high metal binding capacity of metallothioneins (MTs) have also been widely exploited (Cobbett and Goldsbrough 2002).

The MTs encoded genes have been successfully expressed in bacteria and other microorganisms for enhanced heavy metal accumulation. However, one limitation of the application of this methylation-related metal detoxification is that only some metals can be methylated (Ranjard et al. 2003). Numerous microorganisms can mediate the methylation of Pb, Hg, selenium (Se), As, and tin (Sn), which can result in volatilization. These bacteria can transfer a CH_3 group to the metals, resulting in methylated metal compounds, which may differ in volatility, solubility, and toxicity (Gadd 2004; Bolan et al. 2014).

Microbial transformation of metals through oxidation and reduction is also considered an important resistance mechanisms in microbes. For example, microbes can acquire energy through oxidation of Fe, sulfur (S), Mn, and As. On the other hand, microbes during anaerobic respiration can convert metals into a reduced state/ form through dissimilatory reduction. Bacterial species that catalyze such reducing reactions are referred to as dissimilatory metal-reducing bacteria and exploit metals as terminal electron acceptors in anaerobic respiration, even though most of them use Fe^{3+} and S^0 as terminal electron acceptors (Jing et al. 2007). For example, the anaerobic or aerobic reduction of Cr^{6+} to Cr^{3+} by an array of bacterial isolates is an effective means of Cr detoxification (Jing et al. 2007). Also, oxyanions of As, Cr, Se, and U are the terminal electron acceptors used by microbes during anaerobic respiration processes. Moreover, the reduction process performed by microbes is not mainly linked to respiration but to impart metal resistance. Aerobic and anaerobic reduction of Cr^{6+} to Cr^{3+}, Se^{6+} to Se^0, U^{6+} to U^{4+}, and Hg^{2+} to Hg^0 are generally carried out by microbes to detoxify them.

3 Plant Growth-Promoting Rhizobacteria-Assisted Phytoremediation of Environmental Pollutants

3.1 INTRODUCTION

Contamination of soil and water by heavy metals (HMs) and refractory organic pollutants (ROPs) is toxic for living beings and environmental health due to their persistence in nature; low bioavailability; and carcinogenic, mutagenic, and teratogenic nature (Tchounwou et al. 2012; Sharma et al. 2018). Hence, remediation of HMs and ROPs from soil is needed to prevent adverse effects on human health and conserve the environment for future generations. Numerous techniques based on physical and chemical approaches have been implemented for remediation of HMs and ROPs from a contaminated environment, which is viewed as a challenging job with respect to cost and technical complexity (Hashim et al. 2011; Azubuike et al. 2016). In addition, remediation of HM- and ROP-contaminated soil through physical methods has certain limitations, as they alter soil microflora and cause irreversible alterations in soil properties. In the same way, chemical processes for elimination of HMs and ROPs are very expensive, generate secondary pollutants, and produce huge quantities of toxic sludge. These physico-chemical approaches only change the form of the problem and fail to remediate the pollutants thoroughly (Hashim et al. 2011). In this context, application of plants (phytoremediation) with their associated rhizobacterial technique can be the best alternative (Jing et al. 2007; Lugtenberg and Kamilova 2009; Rajkumar et al. 2012). Plant-associated rhizobacteria help in the mitigation of contaminants, as they have plant growth-promoting characteristics along with biocontrol and also help in reduction, removal, or stabilization of toxic contaminants in polluted environments and, ultimately, enhance their phytoremediation efficiency (Sessitsch et al. 2013; Fatima et al. 2018). Thus, utilizing plant growth-promoting rhizobacteria (PGPR) is a new and promising approach for improving the success of phytoremediation in contaminated environments (Rajkumar et al. 2012). This chapter presents an overview of the potential role of PGPR in the bioremediation of organic compounds and heavy metal-polluted environments and provides some discussion of how PGPR could be exploited to enhance phytoremediation.

3.2 PLANT GROWTH-PROMOTING RHIZOBACTERIA

Every plant has its specific microbiome in the rhizosphere, and this microbiome plays a vital role in plant-growth promotion and contaminant detoxification/removal as well (Rajkumar et al. 2012). The bacteria inhabiting the rhizosphere are called rhizobacteria (Glick 2003, 2012). Rhizobacteria present on plant root surfaces help in mineral uptake and contribute to essential vitamins, stomatal regulation, osmotic modification, and adaptation of root morphology. These rhizobacteria bacteria are known as PGPR and have the potential to enhance pollutant tolerance and removal ability of plants and promote plant growth and development by triggering immune responses, modulating the plant hormonal balance, protecting against phytopathogens, and mobilizing nutrients (Glick 2012; Rajkumar et al. 2012; Sessitsch et al. 2013). The term PGPR was coined by Joe Kloepper in the late 1970s and was defined later by Kloepper and Schroth (1978) as "the soil bacteria that colonize the roots of plants by following inoculation on to seed and that enhance plant growth." About 2%–5% of rhizosphere bacteria are PGPR (Antoun and Prevost 2005). PGPR encompass a diverse and large number of bacteria genera, which can be classified according to their activities as: (i) biofertilizers (increasing the bioavailability of nutrients to the plant), (ii) phytostimulators (promoting plant growth and development through the release of phytohormones), (iii) rhizoremediators (breaking down organic contaminants in the rhizosphere), and (iv) biopesticides (protecting the plant from phytopathogens via the synthesis of antibiotics and antifungal metabolites) (Glick 2012). An alternative classification system based on the degree of bacterial proximity to the root and intimacy of association was proposed by Gray and Smith (2005). Accordingly, PGPR can be classified as extracellular PGPR (ePGPR), found in the rhizosphere, on the rhizoplane, or in the spaces between the cells of the root cortex, and intracellular PGPR (iPGPR), which can invade the interior of cells and survive inside, generally in specialized nodular structures. Some well-known examples of ePGPR belong to the genera *Agrobacterium, Arthrobacter, Azotobacter, Azospirillum, Bacillus, Burkholderia, Caulobacter, Chromobacterium, Erwinia, Flavobacterium, Micrococcus, Pseudomonas*, and *Serratia*, whereas bacteria from the genera *Allorhizobium, Azorhizobium, Bradyrhizobium, Mesorhizobium*, and *Rhizobium* belong to the Rhizobiaceae family and are reported as iPGPR (Gray and Smith 2005).

3.2.1 Mechanisms of Plant-Growth Promotion by Rhizobacteria

PGPR promotes plant growth and development in contaminated environments through a variety of direct or indirect mechanisms.

3.2.1.1 Direct Mechanisms

Direct plant-growth promotion is based on either the stipulation of the plants with favorable rhizobacterial compounds or improving the uptake of nutrients by plants from soil (Glick 2012). PGPR promote plant growth and development through a variety of direct mechanisms, including the fixation of atmospheric dinitrogen (N_2); synthesis of siderophores; synthesis and release of different phytohormones or enzymatic activities; and solubilization of inorganic phosphate and mineralization of organic phosphate and/or other nutrients, which then become more readily available for plant growth (Ma et al. 2016a,b). A particular PGPR may affect plant growth and development by using any one or more of these mechanisms. It is probable that the same is true for endophytic bacteria.

3.2.1.1.1 Nitrogen Fixation

Nitrogen (N) is the most essential macronutrient for plant growth and development, and about ~78% of Earth's atmosphere is composed of N. However, N is often a limiting factor for plant growth because atmospheric N exists in gaseous forms, that is, N_2. This N_2 is inaccessible to plants except for a small number of specially adapted prokaryotes, including a few bacteria, cyanobacteria, and actinomycetes. Several bacteria species have the ability to convert (fix) N_2 into plant-available forms, such as ammonia (NH_4^+). These N_2-fixing bacteria can be divided into following two different groups: (i) symbiotic N_2-fixing bacteria: bacteria that infect the roots to produce nodules (e.g., *Allorhizobium, Azorhizobium, Bradyrhizobium, Mesorhizobium,* and *Rhizobium* spp.) and (ii) non-symbiotic N_2-fixing bacteria, also called free-living N_2-fixing bacteria: bacteria that do not penetrate plant tissues but live sufficiently close to the root as to allow the plant to take up the excess of N that is not consumed by its own metabolism (e.g., *Azospirillum* spp. and *Azotobacter* spp.) (van Loon 2007; Beneduzi et al. 2012; Vacheron et al. 2013; Gouda et al. 2018). Generally, they can pass only a small amount of fixed N_2 to the host plants. The process of N_2 fixation is carried out by a complex enzyme known as nitrogenase, an oxygen (O_2)-sensitive enzyme, encoded by *nif* genes that catalyze the reduction of N_2 to NH_4^+. Scientists believe that if *nif* genes are isolated and properly characterized, these genes could be genetically engineered to improve nitrogen fixation (Glick 2012; Bakker et al. 2013). Plants could be genetically engineered to fix their N_2.

3.2.1.1.2 Phosphate Solubilization

Phosphorus (P) is an essential plant macronutrient and plays a key role in plant growth and development. It is the world's second-largest nutritional supplement for crops after N (Rodríguez Fraga 1999). In nature, P is found in organic or inorganic forms that are very poorly soluble; only less than 5% of the total soil P content is available to plants (Sharma et al. 2013). To tackle phosphorus deficit in soil, the application of chemical fertilizers is a common practice. However, chemical fertilizers often comprise elevated concentrations of metals, which can have a detrimental effect on soil microbial life and plants. Thus, phosphate-solubilizing rhizobacteria (PSR) play a key role as biofertilizers because of their ability to supply bioavailable P to plants (Zaidi et al. 2009). PSR may enhance the phytoremediation competence of plants by promoting their growth and health even in the presence of metal stresses. Under metal-stressed conditions, most metal-resistant PGPR can either convert the insoluble P into available forms through acidification, chelation, exchange reactions, and release of organic acids (i.e., lactic, citric, 2-ketogluconic, malic, glycolic, oxalic, malonic, tartaric, valeric, piscidic, succinic, and formic acids) or mineralize organic P by secreting extracellular enzymes such as phosphatase, phytase, phosphonoacetate hydrolase, D-α-glycerophosphatase, and C–P lyase (Rashid et al. 2004; Mahdi et al. 2011; Kalayu

2019). Thus, solubilization and mineralization of phosphorus have been reported by numerous members of bacteria genera like *Azotobacter, Bacillus, Enterobacter, Pseudomonas, Microbacterium, Rhizobium, Serratia, Erwinia, Flavobacterium*, and *Bradyrhizobium* spp. as an important trait in PGPR (Ahemad 2015).

3.2.1.1.3   Iron Sequestration

Iron (Fe) is the fourth most abundant element on the planet and is required for numerous biological processes, which include photosynthesis, respiration, O_2 transport, DNA biosynthesis, gene regulation, and so on. However, it does not normally occur in its biologically relevant ferrous ion (Fe^{2+}) form. Under physiological conditions, Fe can exist in either reduced Fe^{2+} form or oxidized ferric ion (Fe^{3+}) form. In Fe^{3+} form, Fe is not readily available to plants and bacteria and tends to form highly insoluble ferric hydroxides and oxihydroxides. Consequently, the cellular concentration of Fe^{3+} is too low for microorganisms to survive by solely using free Fe. To overcome this Fe nutritional limitation, bacteria, fungi, and some plants synthesize iron-chelating compounds named siderophores. Siderophores are low molecular weight ($\sim$400–1500 Da) organic compounds with a very high and specific affinity to chelate Fe^{3+} as well as having membrane receptors that bind to Fe^{3+}, forming easily assimilable Fe^{3+}-siderophore complexes. Siderophore-producing rhizobacteria, namely *Pseudomonas, Erwinia, Rhizobium, Bacillus, Serratia, Erwinia,* and *Flavobacterium* spp., enhance plant growth by increasing the availability of Fe near the root when they are exposed to elevated concentrations of HMs that unbalance Fe nutrition or by inhibiting the colonization of roots by plant pathogens or other harmful bacteria. Although some plants also have the ability to produce phytosiderophores, these generally present lower affinity for Fe than bacterial siderophores.

3.2.1.1.4   Synthesis and Modulation of Phytohormones

Rhizobacteria synthesis and release of different phytohormones including auxins, cytokinins, and gibberellins, can enhance various stages of plant growth and facilitates the root absorption of HMs in contaminated environment (Glick 2010). Three mechanisms of rhizosphere bacteria involved in heavy metal accumulation in plants are discussed below.

3.2.1.1.4.1   Production of Indole-3-Acetic Acid   Indole-3-acetic acid (IAA) is an active form of auxin produced by many plant-associated bacteria, including PGPR, and controls a wide variety of processes in plant development and plant growth; low concentrations of IAA can stimulate primary root elongation, whereas high IAA levels stimulate the formation of lateral roots, decrease primary root length, and increase root hair formation. The growth-promotion effect of auxin or auxin-like compounds by PGPR may require functional signaling pathways in the host plant. PGPR synthesis of IAA usually occurs through at least five different tryptophan-dependent pathways after its exudation by the roots.

3.2.1.1.4.2   Production of Cytokinins   Cytokinins are a class of phytohormones from which more than 30 typical growth-promoting organic compounds have been described. Production of cytokinins has been well documented in various PGPR such as *Azospirillum, Bacillus, Escherichia, Bradyrhizobium, Arthrobacter, Klebsiella, Proteus, Pseudomonas, Xanthomonas,* and *Paenibacillus.* In plants, many developmental processes are greatly influenced by cytokinins, including enhanced cell division, formation of embryo vasculature, delay of senescence, branching, leaf expansion, chlorophyll production, control root meristem differentiation, nutritional signaling, proliferation of root hairs, and seed germination inducement, but they inhibit lateral root formation and primary root elongation. Inoculation of plants with cytokinin-producing PGPR has been shown to stimulate shoot growth and reduce the root-to-shoot ratio (Arkhipova et al. 2007). Genes responsible for signaling and perception of cytokinin have been identified. These genes represent the positive effect of cytokinin on the entire plant growth. CRE1/AHK4/WOL, AHK2, and AHK3 are the three cytokine receptors (Kakimoto 2003). Expression of the response regulator is activated by these receptors in a cytokinin-dependent manner (Brandstatter and Kieber 1998; Taniguchi et al. 1998). Cytokinin receptors in a bacterial strain of *Bacillus megaterium* rhizobacteria have been found to have a role in plant-growth promotion. A *B. megaterium* UMCV1 strain was initially isolated from the rhizosphere of a *Phaseolus vulgaris* L. plant. The biomass production of two plants, bean and *Arabidopsis thaliana*, was found to be improved by using this bacteria for inoculation (López-Bucio et al. 2007).

3.2.1.1.4.3 Production of Gibberellin Gibberellin, or gibberellic acid, is a large family of well-known tetracyclic diterpenoid phytohormones, entailing up to 136 different structured molecules. From these, 128 have been reported in plants, 7 in fungi, and only 4 in bacteria (GA_1, GA_3, GA_4, and GA_{20}) (Glick 2010, 2012; Kong and Glick 2017). Several findings have revealed that several members of PGPR like *Herbaspirillum seropedicae, Acetobacter, Azospirillum* spp., *Azotobacter* spp., *Bacillus* spp., *Herbaspirillum* spp., *Rhizobium* spp., *Achromobacter xylosoxidans, Acinetobacter calcoaceticus,* and *Gluconacetobacter diazotrophicus* are able to produce gibberellins. Gibberellins also play an important role in plant growth and development, including pollen maturation, leaf and hypocotyl expansion, stem and hypocotyl elongation, seed germination and flowering, lateral root development, root growth, and fruit setting. Gibberellin is well known for its involvement in drought stress. During water stress, increase in gibberellin levels causes closing of stomata, thereby limiting water loss.

3.2.1.1.4.4 Production of 1-Aminocyclopropane-1-Carboxylic Acid Deaminase 1-aminocyclopropane-1-carboxylic acid (ACC) is a precursor molecule in the ethylene synthesis pathway of plants where *S*-adenosyl methionine synthetase converts *L*-methionine to *S*-adenosyl methionine, which is further converted to ACC by the activity of ACC synthetase. ACC oxidase further converts it to ethylene. Ethylene is a phytohormone that can be found in all higher plants and is fundamental to modulate their normal growth and development (Glick 2014). Ethylene inhibits root elongation, and auxin transport promotes senescence and abscission of various organs and leads to fruit ripening. Ethylene is important for their response to stress; therefore, it is known as a stress hormone (Deikman 1997). Under stressed conditions, stress ethylene is synthesized by the plant in response to stresses like

HMs, organic or inorganic chemicals, and physical and biological stresses. Numerous PGPR genera such as *Acinetobacter, Achromobacter, Agrobacterium, Alcaligenes, Azospirillum, Bacillus, Burkholderia, Enterobacter, Pseudomonas, Ralstonia, Pseudomonas* sp., *P. brassicacearum, P. tolaasii, P. fluorescens, P. aeruginosa, P. koreensis, Bacillus* sp., *B. cereus, B. megaterium, Enterobacter* sp., *E. aerogenes, E. intermedius, Arthrobacter* sp., *Firmicutes* sp., *Proteobacteria* sp., *Burkholderia* sp., *Achromobacter xylosoxidans, Actinobacteria* sp., *Methylobacterium oryzae, Serratia* sp., and *Rhizobium* sp. have the capacity to synthesize an enzyme ACC deaminase (encoded by *acdS*) that cleaves ACC into NH_4^+ and α-ketobutyrate, decreasing ethylene levels and subsequently protecting the plant by helping it thrive in heavy metal–contaminated soils and facilitating its growth and development. Thus, ACC deaminase enzyme activity of the PGPR is one property that is helpful in combating ethylene and environmental stress.

3.2.1.1.4.5 Volatile Organic Compounds   Volatile organic compounds (VOCs) are compounds that can enter the atmosphere as vapors due to significantly high vapor pressure. They play a crucial role because their emission has crucial precipitation in many cases of PGPR interaction with plants. The role of VOCs in the bio-control of plant-pathogen and antibiosis is a mechanism that has gained attention in last decade (Vespermann et al. 2007). VOCs emitted by a PGPR, *B. subtilis* GB03, can stimulate many different hormonal signals in *Arabidopsis thaliana*, which include auxin, brassinosteroids, cytokinins, gibberellins, and salicylic acid (Ryu et al. 2003, 2005; Zhang et al. 2007, 2008). This has opened new ways of exploring the role of VOCs in plant growth and development during the relationship between plants and microorganisms.

3.2.1.2   Indirect Mechanisms

PGPR has the capability of indirectly enhancing plant growth through the production and mechanism of bioactive compounds known as allelochemicals, which prevent the deleterious effects of phytopathogens, usually fungi and nematodes. These allelochemicals comprise hydrogen cyanide, antibiotics, and extracellular lytic enzymes, including chitinases, proteases, β-1,3 glucanases, cellulases and laminarinases, and siderophores. Hydrogen cyanide is a volatile compound that exerts moderate biocontrol activity and appears to enhance the effect of bacterial antibiotics when acting synergistically. The production of antibiotics by PGPR, especially from the genera *Bacillus* and *Pseudomonas*, has been comprehensively studied (Glick 2012). The major antibiotics involved in suppression of phytopathogens are nonvolatile compounds, namely pyrrolnitrin, phenazines, phloroglucinols, cyclic lipopeptides, and lipopeptides. As previously mentioned, bacterial siderophores have the ability of chelating Fe^{3+} and forming a soluble Fe^{3+}-siderophore complex. However, in addition to supplying assimilable Fe for plants, PGPR-mediated siderophores also hamper the propagation of phytopathogens, generally fungi, which are incapable of absorbing the chelated Fe complex and are deprived of available Fe in soil. The suppression of phytopathogens is based on antagonism or competition mechanisms and/or by eliciting plant defenses such as induced systemic resistance (ISR). PGPR trigger ISR by different signaling pathways, which include molecules such as ethylene, pyoverdine, N-alkylated benzylamines, flagellar proteins, jasmonate,

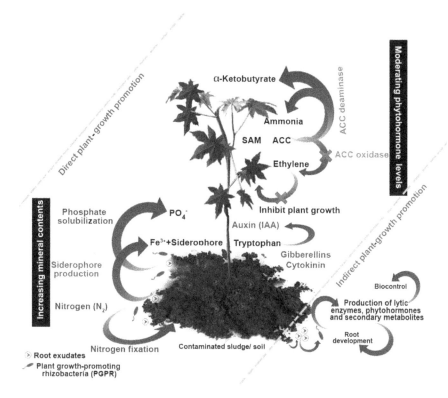

FIGURE 3.1   The mechanisms of plant growth-promoting rhizobacteria (PGPR) in the improvement of plant growth during phytoremediation of contaminated environment. (Modified from Kong Z, Glick BR. 2017. *Advances in Microbial Physiology* 71:97–132.)

chitin, β-glucans, cyclic lipopeptide surfactants, N-acyl-homoserine lactones, and salicylic acid. The antagonistic mechanism of PGPR on phytopathogens is based on the interference of PGPR with virulence factors via effectors delivered by type III secretion systems, whereas in ISR, PGPR bestows the plant's improved resistance against a large array of pathogens. The overall mechanism, including the direct and indirect processes, is presented in Figure 3.1.

3.3   PLANT GROWTH-PROMOTING RHIZOBACTERIA IN PHYTOREMEDIATION OF ENVIRONMENTAL POLLUTANTS

PGPR have the potential to confer plant metal tolerance, enhance biomass production, and prevent phytopathogens from infecting plants, which facilitates the removal of environmental pollutants by plants.

3.3.1   PLANT GROWTH-PROMOTING RHIZOBACTERIA IN PHYTOREMEDIATION OF HEAVY METALS

Accumulation of HMs in the soil environment and their uptake by both PGPR and plants is a matter of growing environmental concern. A variety of plants growing on metalliferous soils have accumulated HMs in their harvestable parts and have the potential to be used for phytoremediation of HM-polluted sites (Etesami 2018). Various species of plants possess different kinds of mechanisms to tolerate HMs, although within one species, more than one mechanism could be in operation. In addition to plant growth, PGPR that assist in the remediation/detoxification of HMs in the phytoremediation process can contribute to this mechanism directly or indirectly. Direct mechanisms involve production of exopolysaccharides (EPS), siderophores, metallothioneins, biosurfactants, organic acids, and phytochelatins, which increases the bioavailability, solubility, and accumulation of HMs and provide a means by which HMs can be removed from solid matrices. On the other hand, indirect mechanisms involve improvement of plant growth through production of phytohormones and protection of plants against pathogens that further facilitate the accumulation of HMs in their tissues. Thus, discoveries of PGPR that are HM resistant and able to promote plant growth in highly polluted environments have raised high hopes for reclamation of heavy metal-polluted soil. Rhizobacteria make excellent biosorbents in the context of their high surface volume and a great number of potentially active chemosorption sites; for instance, teichoic acids and teichuronic acids in the cell wall of Gram-positive bacteria contain active chemical agents with sites capable of passively sequestering HMs. This rhizobacterial immobilization mechanism may lead to the reduction of HM uptake in plants. However, PGPR activity can greatly enhance the mobility of HMs, improve plant performance, and also contribute to plant metal uptake through sorption, mineralization, and transformation mechanisms (de Souza et al. 1999; Whiting et al. 2001; Abou-Shanab et al. 2003). The excretion of EPSs by plant-associated PGPRs prevents the entry of metals into the plant by forming a physical barrier, metal ion exchange, adsorption, precipitation, and complexation with negatively charged functional groups. In addition, bacterial EPS can also lead to the development of soil sheaths around the plant root, which reduces the flow of metals into plants (Dimkpa et al. 2009). By chelating the metal ions, natural organic chelating agents such as biosurfactants, metallophores, organic acid anions, and siderophores (Sessitsch et al. 2013) can decrease metal bioavailability and toxicity (Dimkpa et al. 2008). Examples of some studied successful remediation of heavy metal phytoremediation by using plant growth-promoting rhizobacteria are shown in Table 3.1 (Figure 3.2).

3.3.2 Plant Growth-Promoting Rhizobacteria in Phytoremediation of Organic Pollutants

Phytoremediation of organic pollutants may also occur by phytostabilization or phytostimulation. However, plants have certain limits with respect to their capabilities to remove organic pollutants from the environment (Carvalho et al. 2014). Some organic pollutants can be metabolized by bacteria that may either be found in or added to the soil in the absence of plants; this process is usually slow and inefficient, in part as a consequence of the relatively low number of these degradative bacteria in the soil. Even though these bacteria show high potential to degrade different persistent organic pollutants (POPs), they are unable to survive and proliferate in contaminated soils. The absence of a population of degrader microorganisms can be overcome by the inoculation of the plant rhizosphere with pollutant-degrading strains. Therefore, effective mineralization and degradation of organic pollutants can be

TABLE 3.1

Examples of Some Studies of Successful Remediation of Heavy Metal Phytoremediation by Using Plant Growth-Promoting Rhizobacteria

Bacterial Strain(s)	Host Plant Species	Heavy Metal	Source of Isolation	PGP Traits	Reference
Achromobacter xylosoxidans Ax10	Brassica juncea	Cu	Cu mine soil	ACCD, IAA, P solubilization	Ma et al. (2009b)
Acinetobacter, Alcaligenes, Listeria, Staphylococcus	Nymphaea pubescens	Cu, Zn, Pb	Sediment	–	Kabeer et al. (2014)
Agrobacterium radiobacter	Populus deltoides	Ni	Soil	IAA, siderophores	Wang et al. (2011)
Arthrobacter sp., Pseudomonas aeruginosa, Bacillus licheniformis, Pseudomonas stutzeri	Prosopis juliflora	Cr	Tannery effluent contaminate soil	–	Khan et al. (2014a)
Azospirillum	Panicum virgatum	Pb, Cd	–	–	Arora et al. (2016)
Azotobacter chroococcum, Rhizobium leguminosarum	Zea mays L.	Pb	Soil	IAA production increased and soil pH decreased	Hadi and Bano (2010)
Azotobacter chroococcum HKN-5, Bacillus megaterium HKP-1, Bacillus mucilaginosus HKK-1	Brassica juncea	–	Agronomic soils	P and K solubilization, respectively; metal mobilization	Wu et al. (2006b)
Bacillus cereus N5, Enterobacter cloacae N7	Zea mays L.	Cd, Pb	Soil	IAA, siderophores, ACCD	Abedinzadeh et al. (2019)
Bacillus pumilus, Micrococcus sp.	Noccaea caerulescens	Ni	–	–	Aboudrar (2013)
Bacillus sp. RJ16	Brassica napus	Cd	Soil	IAA	Sheng and Xia (2006)

(Continued)

TABLE 3.1 (Continued)

Examples of Some Studies of Successful Remediation of Heavy Metal Phytoremediation by Using Plant Growth-Promoting Rhizobacteria

Bacterial Strain(s)	Host Plant Species	Heavy Metal	Source of Isolation	PGP Traits	Reference
Bacillus sp. SC2b	Brassica napus	Cd, Pb, Zn	Soil	Ameliorated metal toxicity through biosorption	Ma et al. (2015)
Bacillus subtilis, Bacillus cereus, Flavobacterium sp., Pseudomonas sp.	Orychophragmus violaceus	Zn	Soil	ACCD, IAA, siderophores	He et al. (2010b)
Bacillus thuringiensis GDB-1	Alnus firma	As	Mine tailing waste	ACCD, IAA, siderophores, P-solubilization	Babu et al. (2013)
Bacillus weihenstephanensis SM3	Helianthus annuus	Cu, Zn, Ni	Soil	IAA, P solubilization	Rajkumar et al. (2008)
Bradyrhizobium sp., Rhizobium sp., Sphingomonas sp., Variovorax sp., Burkholderia sp., Janthinobacterium sp., Pseudomonas sp., Streptomyces sp., Nocardia sp., Microbacterium sp., Flavobacterium sp., Pedobacter sp., Chryseobacterium sp., Mucilaginibacter sp.	Salix caprea	Zn, Cd	–	–	Kuffner et al. (2010)
Bradyrhizobium sp., Pseudomonas sp., Ochrobactrum cytisi	Lupinus luteus	Cu, Cd, Pb	Soil	–	Dary et al. (2010)
Brevibacterium casei MH8a	Sinapis alba L.	Cd, Zn, Cu	Soil	IAA, ACCD, hydrocyanic acid production	Płociniczak et al. (2013)
Burkholderia cepacia	Sedum alfredii	Cd, Zn	–	–	Li et al. (2007)

(Continued)

TABLE 3.1 (Continued)

Examples of Some Studies of Successful Remediation of Heavy Metal Phytoremediation by Using Plant Growth-Promoting Rhizobacteria

Bacterial Strain(s)	Host Plant Species	Heavy Metal	Source of Isolation	PGP Traits	Reference
Burkholderia sp. J62	*Lycopersicon esculentum*	Pb, Cd	Paddy field	ACCD, siderophore, IAA, P solubilization	Jiang et al. (2008)
Cellulosimicrobium sp.	*Medicago sativa*	Cr	Soil	IAA, siderophores, P solubilization	Tirry et al. (2018)
Enterobacter ludwigii, Klebsiella pneumoniae, Flavobacterium sp.	*Triticum aestivum*	Hg	—	—	Gontia-Mishra et al. (2016)
Enterobacter sp. JYX7, *Klebsiella* sp. JYX10	*Polygonum pubescens*	Cd	Soil	IAA, siderophores, ACCD, P solubilization	Jing et al. (2014)
Microbacterium oxydans	*Alyssum murale*	Ni	Soil	Ni mobilization	Abou-Shanab et al. (2006)
Microbacterium saperdae, Pseudomonas monteilii, Enterobacter cancerogenus	*Thlaspi caerulescens*	Zn	—	—	Whiting et al. (2001)
Microbacterium sp. CE3R2, *Curtobacterium* sp. NM1R1	*Brassica nigra*	Zn, Pb, Cu, As	Soil	IAA, siderophores, P solubilization	Román-Ponce et al. (2017)
Ochrobactrum sp., *Bacillus* sp.	*Oryza sativa*	Cd, Pb, As	Soil	ACCD, catalase	Pandey et al. (2013)
Paenibacillus macerans NBRFT5, *Bacillus endophyticus* NBRFT4, *Bacillus pumilus* NBRFT9	*Brassica juncea*	Ni	Mix. of fly ash and press mud	Siderophores, organic acids, protons, other nonspecified enzymes	Tiwari et al. (2012)
Pantoea sp. Jp3–3	Guinea grass	Cu	—	—	Huo et al. (2012)
Pseudomonas aeruginosa KUCd1	*Cucurbita pepo*	Cd	Industrial waste-fed canal	Siderophores	Sinha and Mukherjee (2008)

(Continued)

TABLE 3.1 (Continued)

Examples of Some Studies of Successful Remediation of Heavy Metal Phytoremediation by Using Plant Growth-Promoting Rhizobacteria

Bacterial Strain(s)	Host Plant Species	Heavy Metal	Source of Isolation	PGP Traits	Reference
Pseudomonas putida HS – 2	Canola	Ni	Soil	Siderophores, IAA, ACCD	Rodriguez et al. (2008)
Pseudomonas sp. LK9	*Solanum nigrum*	Cd	–	Biosurfactants, siderophores, organic acids	Chen et al. (2014)
Pseudomonas sp. PsA, *Bacillus* sp. Ba32,	*Brassica juncea*	Cr	Soil	ACCD, IAA, siderophores, P solubilization, N fixation	Rajkumar et al. (2006)
Pseudomonas sp. RJ10, *Bacillus* sp. RJ16	*Lycopersicon esculentum*	Cd, Pb	Soil	ACCD, IAA, siderophores	He et al. (2009)
Pseudomonas, *Bacillus*, *Cupriavidus*	*Rinorea bengalensis* (Wall.) O. K., *Dichapetalum gelonioides* sp. *andamanicum* (King) Leenh	Ni	Soil	–	Pal et al. (2007)
Pseudomonas, *Janthinobacterium*, *Serratia*, *Flavobacterium*, *Streptomyces*, *Agromyces*	*Salix caprea*	Zn, Cd	Soil	IAA, ACCD, siderophores	Kuffner et al. (2008)
Psychrobacter sp. SRA1 and SRA2, *Bacillus cereus* SRA10	*Brassica juncea*, *Brassica oxyrrhina*	Ni	Soil	ACCD, siderophores, IAA, P solubilization,	Ma et al. (2009a)

(Continued)

TABLE 3.1 (Continued)

Examples of Some Studies of Successful Remediation of Heavy Metal Phytoremediation by Using Plant Growth-Promoting Rhizobacteria

Bacterial Strain(s)	Host Plant Species	Heavy Metal	Source of Isolation	PGP Traits	Reference
Ralstonia sp. TISTR 2219, *Arthrobacter* sp., TISTR 2220	*Ocimum gratissimum*	Cd	Soil	—	Prapagdee and Khonsue (2015)
Sinorhizobium, *Agrobacterium*	*Medicago lupulina*	Cu, Zn	Soil	—	Jian et al. (2019)
Sphingomonas macrogoltabidus, *Microbacterium liquefaciens*, *Microbacterium arabinogalactanolyticum*	*Alyssum murale*	Ni	—	—	Abou-Shanab et al. (2003)
Staphylococcus arlettae NBRIEA G-6	*Brassica juncea*	As	Soil	IAA, siderophores, ACCD	Srivastava et al. (2013)

Abbreviations: P, Phosphate; IAA, Indole acetic acid; ACCD, 1-Aminocyclopropane-1-carboxylic acid deaminase; PGP, Plant growth-promoting.

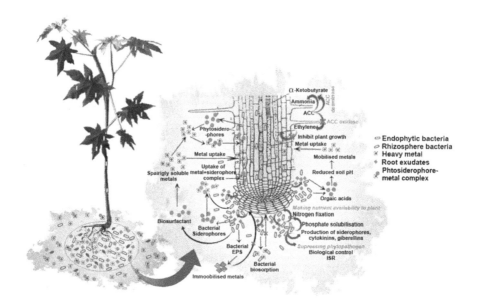

FIGURE 3.2 Role of plant growth-promoting rhizobacteria bacteria in alleviation of heavy metal toxicity, phytoextraction, and phytostabilization. (Reprinted with permission from Kumar and Chandra 2020a. In: Saxena G, Bharagava RN (eds) *Bioremediation of Industrial Waste for Environmental Safety. Vol II—Microbes and Methods for Industrial Waste Management.* Singapore: Springer.)

achieved by the combined use of plants and their root-associated rhizosphere bacteria (Glick 2010). Several plants growing in polluted soil and water that host different types of rhizosphere bacteria able to degrade organic pollutants have been isolated, and degradation pathways and genes involved in organic pollutant degradation have been identified (Nicoară et al. 2014; Fatima et al. 2015). Soil microbes assist host plants in different reactions and metabolic processes that occur in the biogeochemical cycling of nutrients, maintenance of soil structure, and detoxification of pollutants (Khan et al. 2013). In such a relationship, rhizosphere bacteria with catabolic genes feed upon the organic pollutants as a sole source of carbon for their metabolism and cell functioning, whereas plants facilitate the survival of bacteria by adjusting the rhizosphere environment through production of root exudates, rhizosphere oxidation, co-metabolite induction, H^+/OH^- ion excretion, organic acid production, and release of biogenic surfactants. Plant–rhizosphere bacteria interactions enhance the abundance and expression of catabolic genes in the rhizosphere, leading to an increase in mineralization, degradation, stabilization, or sequestration of a variety of organic pollutants (Passatore et al. 2014; Jha and Jha 2015). Recently, several studies have been conducted to explore the potential of plant–rhizosphere bacteria partnerships for the remediation of organic pollutant–contaminated soil and water (Qin et al. 2014; Jha and Jha 2015). Several microorganisms have diverse metabolic capabilities to degrade and utilize petroleum hydrocarbons as their carbon and energy sources via aerobic or anaerobic pathways. Many other studies have also reported enhanced degradation of organic pollutants by plant–rhizosphere bacteria partnerships, as shown in Table 3.2.

TABLE 3.2
Example of Successful Degradation of Organic Pollutants by Plant–Rhizosphere Bacteria Partnership

Bacteria Species	Host Plant Species	Targeted Pollutant(s)	Reference
Rhodococcus, Luteibacter, Williamsia	*Salix caprea, Pinus nigra*	PCBs	Leigh et al. (2006)
Achromobacter sp.	*Nicotiana tabacum*	PCBs	Ionescu et al. (2009)
Azospirillum lipoferum spp.	*Triticum* sp.	Crude oil	Muratova et al. (2005)
Bacillus pumilus	*Armoracia rusticana*	PCBs	Ionescu et al. (2009)
Burkholderia cepacia	*Hordeum sativum* L.	2,4-D	Jacobsen (1997)
Exiguobacterium aestuarii strain ZaK	*Zinnia angustifolia*	RBB dye	Khandare et al. (2012)
Microbacterium foliorum, Gordonia alkanivorans, Mesorhizobium	*Sesbania cannabina*	TPH	Maqbool et al. (2012)
Microbacterium oxydans type strain	*Pinus nigra*	PCBs	Siciliano and Germida (1998a)
Pseudomonas	*Hordeum sativum* L.	Phenanthrene	Ankohina et al. (2004)
Pseudomonas aeruginosa strain R75, *Pseudomonas savastanoi* strain CB35	*Elymus dauricus*	CBA	Siciliano and Germida (1998a)
Pseudomonas fluorescens	*Medicago sativa*	PCBs	Brazil et al. (1995)
Pseudomonas fluorescens	*Triticum* sp.	TCE	Yee et al. (1998)
Pseudomonas fluorescens strain F113	*Medicago sativa*	PCBs	Villacieros et al. (2005a)
Pseudomonas mendocina, Pseudomonas fluorescens	*Solanum nigrum*	PCBs	Ionescu et al. (2009)
Pseudomonas putida Flav1-1, *Pseudomonas putida* PML2	*Arabidopsis*	PCBs	Narasimhan et al. (2003)
Pseudomonas putida PCL1444	*Lolium multiflorum*	Naphthalene	Kuiper et al. (2004)
Pseudomonas, Rhodococcus, Rhizobium,	*Medicago sativa*	PCBs	Ionescu et al. (2009)
Sinorhizobium meliloti strain A-025	*Medicago sativa*	Aroclor compounds	Mehmannavaz et al. (2002)
Sphingobacterium mizutae, Burkholderia cepacia	*Salix caprea*	PCBs	Ionescu et al. (2009)
Sphingomonas herbicidovorans, AB042233, *Sphingomonas* sp. DS3-1, *Sphingomonas taejonensis, Sphingomonas* sp. D12	*Zea mays*	$\alpha, \beta, \gamma, \delta$-HCH	Abhilash et al. (2013) Böltner et al. (2008)

Abbreviations: PCBs, Polychlorinated biphenyls; TPH, Total petroleum hydrocarbon; HCH, Hexachlorocyclohexane; 2,4-D, 2,4-dichlorophenoxyacetic acid; RBB, Remazol Black B; CBA, Chlorinated benzoic acid; TCE, Trichloroethylene.

4 Endophyte-Assisted Phytoremediation of Organic and Inorganic Pollutants

4.1 INTRODUCTION

Phytoremediation is a promising green remediation technology where plants are grown in the presence of contaminants to enhance the decomposition or removal of inorganic and organic environmental contaminants (Cunningham et al. 1995; Ali et al. 2013; Ashraf et al. 2019). There have been many studies and reports on the successful use of phytoremediation technology for the clean-up of sites contaminated with volatile or non-volatile organic pollutants, heavy metals (HMs), pesticides, and so on (McGrath and Zhao 2003; Grobelak and Napora 2015; Suman et al. 2018). In the case of phytoremediation of organic pollutants in soil, plants are used to take up or enhance the decomposition of pollutants into non- or less-toxic forms (Cunningham and Berti 1993; Newman and Reynolds 2004; Khan et al. 2014a,b; Zazouli et al. 2014). Additionally, plants can improve the soil structure by increasing aeration and humidity and also promote microbial growth (Narasimhan et al. 2003; Kuiper et al. 2004; Chandra et al. 2018a,b). Although plants and their associated microorganisms can degrade a wide range of organic xenobiotics within the plant, the rhizosphere, or both, many compounds are degraded either slowly or not at all (Germaine et al. 2006, 2009). However, successful field phytoremediation has been limited due to the high phytotoxicity of organic compounds, causing growth inhibition, reduced transpiration, chlorosis, and wilting (Barac et al. 2004). Microbes that reside in the inner tissues of living plants without causing apparent negative symptoms of infection have been termed endophytes and have been demonstrated to play a key role in host plant adaptation in polluted environments (Doty 2008; Huo et al. 2012). In plant–endophyte symbiosis, endophytes receive carbohydrates from plants and, in return, can improve plants' resistance to external stresses such as contaminants, temperature extremes, water and nutrient limitations, salt, and pathogens (Weyens et al. 2009a,b; Ma et al. 2016a,b). Thus, the use of endophytes capable of degrading and detoxifying toxic organic compounds in combination with specific plants (chosen to suit the environment to be remediated) could offer an efficient and sustainable remediation technology for the twenty-first century (Weyens et al. 2010c,d; van der Lelie et al. 2009). Plants and their associated endophytic bacteria interact with each other for mutual benefits, and these interactions can be exploited to enhance the remediation of polluted soil and water. This approach, which is simple in both concept and execution, may provide workers with a straightforward, but highly effective, method to facilitate the phytoremediation of a range of organic contaminants (Luo et al. 2011, 2012; Chen et al. 2014; Ashraf et al. 2018a,b).

4.2   WHAT ARE ENDOPHYTES?

Bacteria, fungi, and actinomycetes on plant roots and in the rhizosphere benefit from root exudates, but some bacteria, fungi, and actinomycetes are capable of entering plant tissues as endophytes that do not cause harm and could establish a mutualistic association (Weyens et al. 2009a,d; Wu et al. 2009; Ma et al. 2016b). The German botanist Heinrich Friedrich Link was the first to describe endophytes in 1809 as a distinct group of partly parasitic fungi living in plants. The most common definition of endophytes is derived from the practical description given in 1997 by Hallmann and coauthors, who stated that endophytes are those (fungi, bacteria, or actinomycetes) that reside internally in plant tissues, can be isolated from the plant after surface disinfection, and cause no negative effects on plant growth. Recent molecular advances require that this definition be adjusted since an abundance of unculturable endophytes has been sequenced. It appears that certain fungal endophytes can shift between parasitic and mutualistic life strategies, described as a balanced antagonism. Hence, a more suitable definition of endophytes is the set of microbial genomes located inside plant organs. They have established harmonious associations with host plants including symbiotic, mutualistic, commensalistic, and trophobiotic relationships over a long evolutionary process. In accordance with their plant-inhabiting life strategies, endophytes can be either "obligate" or "facultative." Obligate endophytes are unable to proliferate outside of plants and are likely transmitted via seed rather than originating from the rhizosphere, whereas facultative endophytes have a stage in their life cycle in which they exist outside host plants. However, they will colonize plants through coordinated infection when the opportunity arises.

4.3   MICROBIAL COLONIZATION

Microbial colonization of the rhizoplane and/or root tissues is known as root colonization. There are several routes for endophytes to colonize within plants, as described by Hallmann (2001) and James et al. (2002). Several rhizospheric bacteria do not only colonize the rhizosphere and/or the rhizoplane but can also enter plants and colonize internal tissues (roots and stems), and many of them have shown plant growth-promoting effects. Root endophytes often colonize and penetrate the epidermis at sites of lateral root emergence, below the root hair zone, and in root cracks. These colonizers are capable of establishing populations both inter- and intracellularly. After penetration of bacteria with plan root endodermis, the xylem vascular system is the major transport route for systemic colonization of internal plant compartments. In general, endophytes enter plant tissue through the flower, leaf, stem, root-zone, and cotyledon, and they may either become localized at the point of entry or spread throughout the plant (Rajkumar et al. 2009). Several recent studies confirm that plants host diverse endophytic communities and that endophytic bacteria mostly derive from the rhizosphere. However, phyllosphere bacteria may also be a source of endophytes (Hallmann et al. 1997). Depending on the strain, numerous colonization routes, including passive or active mechanisms enabling bacteria to migrate from the rhizoplane to the cortical cell layer, where the plant endodermis represents a barrier for further colonization, have been described. But how do these bacteria initially enter their hosts? The best evidence suggests that they most likely enter at lateral root junctions, naturally occurring cracks. However, it should be stressed that this is unlikely to be an entirely passive process, as many endophytic bacteria express cell wall–degrading enzymes (CWDEs), albeit generally in lower concentrations than expressed by plant pathogens. It has been proposed that cellulolytic and pectinolytic enzymes produced by endophytes are involved in the infection process (Hallmann et al. 1997), as in *Klebsiella* strains, pectate lyase has been implicated in participating during plant colonization (Kovtunovych et al. 1999). The CWDEs such as endogluconase and polygalacturonase seem to be required for the infection of *Vitis vinifera* L. by *Burkholderia* sp. (Compant et al. 2005). Also, a few bacteria (e.g., some *Herbaspirillum* spp.) have been shown to possess T3SSs, which are the route of exit for excreted plant CWDEs, although most do not. Other, more passive modes of entry are often through natural breaks in roots or root tips and/or by vegetative propagation. Bacteria might even migrate to reproductive organs of Angiospermae plants and have also been detected in the internal tissues of seeds (tegument), fruits (pulp), and flowers (epidermis and ovary) and in pumpkin flowers, as well as in the pollen of pine. Endophyte numbers are generally lower in aerial parts than in roots, which suggests that although there is some upward movement of endophytes within their hosts, perhaps through the transpiration stream, this movement is limited and may only be possible for bacteria that express CWDEs and/or T3SSs. Pollutant-degrading endophytic bacteria may show preferential colonization for specific plant tissue. For instance, alkane-degrading root endophytic bacteria showed better colonization in the root, whereas shoot endophytic bacteria showed better colonization in the shoot (Andria et al. 2009; Yousaf et al. 2011). However, higher numbers of pollutant-degrading endophytic bacteria were observed in roots of most plants as compared to shoots and leaves (Rosenblueth and Martínez-Romero 2006). Moreover, lower densities of alkane-degrading endophytic bacteria were observed at the initial stages of plant growth, indicating that first bacteria have to establish themselves in the plant environment and later become important for the degradation of organic pollutants (Afzal et al. 2011). Bacterial colonization and population inside plant tissues can be monitored by labeling inoculant strains with markers, such as the *gusA* and *gfp* genes. The colonization of large numbers of green fluorescent protein-labeled endophytic bacteria was studied in the roots and shoots of poplar and pea plants (Germaine et al. 2004). Successful colonization by endophytes depends on many variables, including plant tissue type, plant genotype, the microbial taxon and strain type, and biotic and abiotic environmental conditions (Hardoim et al. 2008). Conrath et al. (2006) suggested that colonization of microbes involves various factors such as plant genotype, growth stage, physiological status, type of plant tissues, and some soil environmental conditions, as well as some agricultural practices. However, the microbial metabolic pathways of colonization may play a significant role as determinants of endophytic diversity (Compant et al. 2010; Philippot et al. 2013). Bacterial endophytes can enter into roots via colonization of root hairs (Mercado-Blanco and Prieto 2012), whereas, in a few cases, stem and leaf also produce exudates that attract microorganisms (Hallmann 2001). Hence, only adapted bacteria can exist and enter the plant via stomata, wounds, and hydathodes.

Besides bacterial endophyte colonization, different colonization strategies have been described for clavicipitaceous and nonclavicipitaceous fungal endophytes. Species of *Clavicipitaceae*, including *Balansia* spp., *Epichloë* spp., and *Claviceps* spp., establish symbioses almost exclusively with grass, rush, and sledge hosts, in which they may colonize the entire host plant systemically. They proliferate in the shoot meristem, colonizing intercellular spaces of the newly forming shoots, and can be transmitted vertically via seeds. Some *Neotyphodium* and *Epichloë* species may also be transmitted horizontally via leaf fragments falling on the soil. Based on colonization characteristics, Rodrigues et al. (2017) classified clavicipitaceous endophytes as class I fungal endophytes. Fungi colonizing above- and below-ground plant tissues and being horizontally and/or vertically transmitted were grouped as class II fungal endophytes. Class III endophytes were defined to contain mostly members of Dikaryomycota (Ascomycota or Basidiomycota), which are particularly well studied in trees but also in other plant taxa and in various ecosystems Members of this class are mostly restricted to aerial tissues of various hosts and are horizontally transmitted. Class IV endophytes comprise dark, septate endophytes, which, similar to mycorrhizal fungi, are restricted to roots, where they reside inter- and/or intracellularly in the cortical cell layer. In general, endophytes are more likely to show plant growth-promoting effects than bacteria exclusively colonizing the rhizosphere.

4.4   ENDOPHYTIC BACTERIA AND THEIR ECOLOGY

Endophytic bacteria can be defined as those bacteria that reside within the interior tissues of plants without showing external sign of infection or negative effects on their host. Because endophytic bacteria can proliferate within plant tissue, they are likely to interact closely with their host and therefore face less competition for nutrients and are more protected from adverse changes in the environment than bacteria in the rhizosphere and phyllosphere. Endophytic bacteria in plants were first reported by Darnel in 1904. They have since been reported to be associated with almost every plant species, and of the nearly 300 000 plant species that exist on the earth, each individual plant is host to one or more endophytes. Only a few of these plants have ever been completely studied relative to their endophytic biology. It has been reported that about 10^5 cfu of endophytic bacteria are present per gram of fresh root weight, and their diversity so large that 70%–80% of them have yet to be identified, despite recent advances. However, some other studies have also reported that endophytes can originate from the phyllosphere. Endophytic bacteria can be facultative, obligate, or passive, depending on the genotype of the host plant and life strategy. Facultative endophytes are free living and present in the soil and colonize plants opportunistically. They are also the species most commonly cultivated from plants. Facultative endophytic bacteria can survive and colonize outside the plant during a period of their life cycle. However, obligate endophytic bacteria depend on the host plant for their survival and metabolic activities and may be transmitted from one generation to the next through seeds or vegetative plant tissues. The third group, the passive endophytes, do not actively seek to colonize the plant but do so as a result of stochastic events, such as an open wound along root hairs. This passive life strategy may cause the endophytes to be less competitive since the cellular machinery required for plant colonization is lacking and therefore may be less appropriate as plant growth promoters.

Endophytic populations, like rhizospheric populations, are conditioned by biotic and abiotic factors, but endophytic bacteria could be better protected from biotic and abiotic stresses than rhizospheric bacteria. Moreover, many endophytic bacteria, particularly those inhabiting plants growing in a polluted environment, produce degradative enzymes and contribute to the degradation of several types of organic compounds present in the rhizosphere and endosphere. Some endophytes have no apparent effects on plant performance but live on the metabolites produced by the host. These are termed commensal endophytes, whereas other endophytes confer beneficial effects to the plant, such as protection against invading pathogens and (arthropod) herbivores, either via antibiosis or induced resistance, and plant-growth promotion. Generally, endophytes can have neutral or detrimental effects on the host plant under normal growth conditions, whereas they can be beneficial under more extreme conditions or during different stages of the plant life cycle. It has not been resolved whether plants benefit more from an endophyte than from a rhizospheric bacterium. It is still not always clear which population of microorganisms (endophytes or rhizospheric bacteria) promotes plant growth; nevertheless, benefits conferred by endophytes are well recognized and will be presented here.

Endophytic bacteria have been isolated from both monocotyledonous and dicotyledonous plants, ranging from woody tree species like oak and pear, to herbaceous crop plants such as sugar beet and maize. The most common endophytes are typed as commensals, with unknown or yet unknown functions in plants, and less common ones are those shown to have positive (mutualistic) or negative (antagonistic) effects on plants. Most of our knowledge about endophytic bacteria comes from work on a few well-studied "model" organisms, such as *Azoarcus*, *Burkholderia*, *Gluconacetobacter*, *Herbaspirillum*, and *Klebsiella* spp., which were all isolated from non-legumes, particularly grasses. Endophytic bacteria have been widely studied in various climatic and geographic zones and were found to be ubiquitous within all examined plants to date. In general, bacteria in the family Pseudomonaceae, Burkholderiaceae, and Enterobacteriaceae are the most frequently cultivable endophytes found in plant tissues, although recent biotechnological advances have exposed the presence of additional taxa, including non-culturable endophytes. Several studies revealed that endophytic bacteria mainly reside in the intercellular apoplast and in dead or dying cells. They are also found in xylem vessels, within which they may be translocated from the root to the aerial parts. It has been reported that a significant number of endophytes (10^3–10^6 cells) can colonize the plant vascular system. The highest densities of endophytic bacteria usually are observed in the roots and decrease progressively from the stem to the leaves.

Endophytic bacteria, which have been found in numerous plant species, often belong to genera commonly found in soil, including *Pseudomonas*, *Burkholderia*,

Bacillus, and *Azospirillum*. In addition, most of the endophytic bacteria (26%) could be assigned to the ϒ-proteobacteria, including 56 recognized and other unidentified genera as well as the *Candidatus portiera*genus. Endophytic ϒ-proteobacteria are mostly represented by a few genera, namely *Pseudomonas, Enterobacter, Pantoea, Stenotrophomonas, Acinetobacter*, and *Serratia* (>50 sequences each). Among Gram-positive endophytic bacteria, the class Actinobacteria (20%) comprises diverse endophytes belonging to 107 recognized and 15 unidentified genera. Diversity associated with bacterial endophytes exists, not only in the plant species colonized but also in the colonizing bacterial taxa. Plants can harbor dozens of symbiotic bacterial species within roots and stems, and this microbial community can be altered according to the environmental conditions. For example, plants growing in a petroleum-contaminated soil showed a preference for petroleum-degrading bacteria in the root interior, and this preference was plant species specific (Siciliano et al. 2001). Endophytic bacterial species isolated from plants, to date, include Gram-positive and Gram-negative bacteria and include 82 genera, such as *Acidovorax, Acinetobacter, Actinomyces, Aeromonas, Afipia, Agrobacterium, Agromonas, Alcaligenes, Alcanivorax, Allorhizobium, Alteromonas, Aminobacter, Aquaspirillum, Arthrobacter, Aureobacterium, Azoarcus, Azomonas, Azorhizobium, Azotobacter, Azospirillum, Bacillus, Beijerinckia, Blastobacter, Blastomonas, Brachymonas, Bradyrhizobium, Brenneria, Brevundimonas, Burkholderia, Chelatobacter, Chromobacterium, Chryseomonas, Comamonas, Corynebacterium, Delftia, Derxia, Devosia, Enterobacter, Flavimonas, Flavobacterium, Flexibacter, Frankia, Halomonas, Herbaspirillum, Matsuebacter, Mesorhizobium, Moraxella, Nevskia, Nocardia, Ochrobactrum, Pantoea, Pectobacterium, Phenylobacterium, Phyllobacterium, Photobacterium, Porphyrobacter, Pseudoalteromonas, Pseudomonas, Psychrobacter, Ralstonia, Renibacterium, Rhizobacter, Rhizobium, Rhizomonas, Rhodanobacter, Rhodococcus, Shewanella, Sinorhizobium, Sphingobacterium, Sphingomonas, Spirillum, Stenotrophomonas, Streptomyces, Thauera, Variovorax, Vibrio, Xanthomonas, Xylella, Zoogloea, Zymobacter, Zymomonas*, and *Methylobacterium*. These endophytic bacteria are widely distributed within α-proteobacteria, β-proteobacteria, ϒ-proteobacteria, Firmicutes, Actinobacteria, and Bacteroidetes, and most of the endophytic bacteria belong to α-proteobacteria, β-proteobacteria, and ϒ-proteobacteria. Endophyte colonization has also been visualized with the use of the β-glucuronidase (GUS) reporter system. A GUS-marked strain of *Herbaspirillum seropedicae* Z67 was inoculated onto rice seedlings. GUS staining was most intense on coleoptiles, lateral roots, and also at some of the junctions of the main and lateral roots (James et al. 2002). This study by James et al. (2002) showed that endophytes entered the roots through cracks at the point of lateral root emergence. *Herbaspirillum seropedicae* subsequently colonized the root intercellular spaces, aerenchyma, and cortical cells, with a few penetrating the stele to enter the vascular tissue. The xylem vessels in leaves and stems were also colonized. The use of molecular techniques to identify microorganisms is currently a key tool to study endosphere ecology. Surface sterilization of the root followed by tissue maceration is common to ensure isolation of endophytes from other rhizosphere and rhizoplane bacteria followed by characterization by a number of techniques, such as fatty acid methyl ester (FAME) analysis and morphological or enzymatic tests. Despite the long history of culturing microorganisms, it is clear that most are unculturable in the laboratory setting. Culture-independent techniques based on polymerase chain reaction (PCR) are followed by downstream methods to analyze the endophyte community, which may include cloning or other community fingerprinting techniques such as restriction fragment length polymorphism (RFLP), terminal restriction fragment length polymorphism (T-RFLP), denaturing gradient gel electrophoresis (DGGE), and amplified rDNA restriction analysis (ARDRA). With the availability of broader meta-approaches, discovering and characterizing the endophyte community structure and dynamics in its entirety is increasingly achievable. Recently, endophyte research has been introduced to the expanding array of next-generation technologies in an effort to probe the vast amount of information in endophyte genomes, proteomes, and transcriptomes. More recently, microarrays were used to characterize transcriptomes of endophytes to identify genes that are up-regulated in the presence of host plants (Matilla et al. 2007; Shidore et al. 2012). These types of studies are helping us understand the genes involved in the intricately complex plant–microbe interactions. High-throughput sequencing of 16S rRNA has recently been used to define the core endophytic bacterial microbiome of *A. thaliana*.

In comparison with rhizosphere and phyllosphere bacteria, endophytic bacteria are likely to interact more closely with their host. In these very close plant–endophyte interactions, plants provide nutrients and residency for bacteria, which in exchange can directly or indirectly improve plant growth and health. Endophytes perform important ecological functions during the plant's life. The most important is protection against pathogens, interaction with plant symbionts, eliciting plant defense mechanisms against environmental stresses, production of volatile substances, and nitrogen fixation.

4.5   PLANT-GROWTH PROMOTION BY ENDOPHYTIC BACTERIA

The beneficial effects of bacterial endophytes on their host plant appear to occur through similar mechanisms as described for plant growth-promoting rhizobacteria (PGPR) in Chapter 3. Endophytic bacteria can improve plant growth and development in a direct or indirect way (Santoyo et al. 2016). In addition to their beneficial effects on plant growth, endophytes have considerable biotechnological potential to improve the applicability and efficiency of phytoremediation.

4.5.1   DIRECT PLANT GROWTH-PROMOTING ACTIVITY

Direct plant growth-promoting mechanisms of endophytic bacteria may involve biological fixation of atmospheric nitrogen (diazotrophy) and supplying it to plants; the production of small signaling growth regulators (phytohormones) such as indole-3-acetic acid (IAA), auxins, cytokinins, and gibberellin that can act to enhance various stages of plant growth; suppression of the production of stress ethylene by 1-aminocyclopropane-1-carboxylate (ACC) deaminase activity; and alteration of sugar-sensing mechanisms in plants. They synthesize siderophores that can solubilize and sequester iron from the soil and provide it to the plant. They may have mechanisms for the solubilization of mineral nutrients such as phosphorus and potassium that will become more available for plant growth, and they may synthesize some less well-characterized, low-molecular-mass compounds or enzymes that can modulate plant growth and development.

4.5.2 INDIRECT PLANT GROWTH-PROMOTING ACTIVITY

Endophytic bacteria can also indirectly benefit plant growth by preventing the growth or activity of phytopathogens (i.e., nematodes, insects, pests, fungi) through competition for space and nutrients, antibiosis, production of hydrolytic enzymes, and inhibition of pathogen-produced enzymes or by inducing a systemic resistance of plants against pathogens. These mechanisms are briefly outlined in Chapter 3. It is believed that certain endophyte bacteria trigger a plant defense reaction known as induced systemic resistance (ISR), leading to a higher tolerance of pathogens. ISR is effective against different types of pathogens, but the inducing bacterium does not cause visible symptoms on the host plant. Bacterial strains of the genera *Pseudomonas* and *Bacillus* can be considered the most common groups inducing ISR, although ISR induction is not exclusive to these groups. Bacterial factors responsible for ISR induction were identified to include flagella, antibiotics, N-acylhomoserine lactones, salicylic acid, jasmonic acid, siderophores, volatiles (e.g., acetoin), and lipopolysaccharides. The shoot endophyte *Methylobacterium* sp. strain IMBG290 was shown to induce resistance against the pathogen *Pectobacterium atrosepticum* in potato in an inoculum-density-dependent manner. The observed resistance was accompanied by changes in the structure of the innate endophytic community. Endophytic community changes were shown to correlate with disease resistance, indicating that the endophytic community as a whole, or just fractions thereof, can play a role in disease suppression. Bacterial endophytes also produce antimicrobial compounds. For example, the endophyte *Enterobacter* sp. strain 638 produces antibiotic substances, including 2-phenylethanol and 4-hydroxybenzoate. Generally, endophytic actinomycetes are the best-known examples of antimicrobial compound producers, and compounds discovered so far include munumbicins, kakadumycins, and coronamycins. Some of these compounds appear to be valuable for clinical or agricultural purposes, but their exact roles in plant–microbe interactions still need to be elucidated.

4.6 ROLE OF SECONDARY METABOLITES
OF ENDOPHYTIC BACTERIA

Secondary metabolites are biologically active organic compounds that are an important source of antifungal, anti-oomycete, antibacterial, anticancer, antioxidant, antidiabetic, immunosuppressive, insecticidal, nematicidal, and antiviral agents. Many endophytic members of common soil bacterial genera, such as *Pseudomonas*, *Burkholderia*, and *Bacillus* are well known for their diverse range of secondary metabolic products, including antibiotics; anticancer compounds; volatile organic compounds; and antifungal, antiviral, insecticidal, and immunosuppressant agents. Moreover, endophyte metabolites are involved in mechanisms of signaling, defense, and genetic regulation of the establishment of symbiosis. Besides the production of secondary metabolite compounds, endophytes are also able to influence the secondary metabolism of their host plant. This was demonstrated in strawberry plants inoculated with a *Methylobacterium* species strain, where the inoculant strain influenced the biosynthesis of furanones, a flavor compound, in the host plants. Biosynthesis and accumulation of phenolic acids, flavan-3-ols, and oligomeric proanthocyanidins in bilberry (*Vaccinium myrtillus* L.) plants were enhanced upon interaction with a fungal endophyte, a *Paraphaeosphaeria* sp. strain. One member of the plant-associated fluorescent *Pseudomonas viridiflava*, which has been isolated on and within the tissues of many grass species (Miller et al. 1998), was found to produce two novel antimicrobial compounds called ecomycins. It was found that these compounds were able to inhibit the human pathogens *Candida albicans* and *Cryptococcus neoformans*. Bioplastics production from endophytic bacteria is receiving increasing commercial interest. Lemoigne (1926) first described a bioplastic, poly-3-hydroxybutyrate (PHB) produced by *Bacillus megaterium*. Genomic analysis indicates that many species of bacteria have the potential to produce bioplastics (Kalia et al. 2003).

4.7 PLANT–ENDOPHYTE RELATIONSHIPS IN PHYTOREMEDIATION

Endophyte-assisted phytoremediation can also be applied for the clean-up of soil polluted with both organic compounds and heavy metals. It is a promising approach to improve plant biomass production and phytoremediation of co-contamination of toxic metals and organic pollutants. There are several advantages associated with the use of endophytic bacteria in phytoremediation of contaminated environments. A major advantage of using endophytic bacteria over rhizospheric bacteria in phytoremediation is that rhizospheric bacterial population is difficult to control, and competition between rhizospheric bacterial strains often reduces the number of the desired strain. The toxic pollutants taken up by the plant may be degraded *in planta* by endophytic degraders, reducing the toxic effects of contaminants in environmental soil on flora and fauna. The use of endophytes that naturally inhabit the internal tissues of plants reduces the problem of competition between bacterial strains. Furthermore, their population can be monitored by sampling plant tissue. Studies suggest that these bacteria can be used to complement the metabolic potential of their host plant through direct degradation as well as the transfer of degradative plasmids to other endophytes. In addition to pollutant degradation pathways, endophytic bacteria may also possess the capability to enhance plant growth and adaptation in polluted environments through their direct and indirect plant growth-promoting mechanisms.

4.7.1 ENDOPHYTIC BACTERIA IN PHYTOREMEDIATION OF HEAVY METALS

Endophytes play a vital role in phytoremediation of heavy metals. During phytoremediation of metal-contaminated sites, heavy metal resistance or sequestration systems of bacterial endophytes lower metal phytotoxicity in hosts and increase metal accumulation and translocation to above-ground plant tissues; thus, they play a significant role in the adaptation of plants to a polluted environment (Sessitsch et al. 2013). Additionally, they also change metal bioaccumulation ability in plants by excreting metal-immobilizing extracellular polymeric substances (EPSs), as well as metal-mobilizing organic acids and biosurfactants (Marchut-Mikolajczyk et al. 2018). The EPSs secreted by endophytic bacteria, consisting mainly of polysaccharides, proteins, nucleic acids, and lipids, also play a significant role in metal complexation, thereby reducing their bioaccessibility and bioavailability. Endophytic bacteria isolated from metal-hyperaccumulating plants exhibit tolerance to high metal concentrations (Idris et al. 2004). This may be due to the presence of high concentrations of heavy metals in hyperaccumulators, modulating endophytes to resist/adapt to such environmental conditions. It is also possible that metal-hyperaccumulating plants may simultaneously be colonized by different metal-resistant endophytic bacteria in a wide variety of Gram-positive and Gram-negative bacteria (Rajkumar et al. 2009). Metal-resistant endophytic bacteria that also produce siderophores have been isolated from many different plant species and promote plant growth in elevated concentrations of heavy metals. Barzanti et al. (2007) reported that 81% of bacterial isolates recovered from *Alyssum bertolonii* were shown to produce siderophores and to promote plant growth under Ni stress. Bacterial endophytes might function more effectively than bacteria added to the soil because they participate in a process known as *bioaugmentation* (Newman and Reynol 2005). Table 4.1 summarizes the role of endophytic bacteria in heavy metal remediation of contaminated soil.

4.7.2 ENDOPHYTIC BACTERIA-ASSISTED PHYTOREMEDIATION OF ORGANIC POLLUTANTS

The presence of recalcitrant organic pollutants in the environment inhibits the growth and metabolic activities of plants and microorganisms even at very low concentrations. Thus, phytoremediation of these toxic organic pollutants is often inefficient because plants do not completely degrade these compounds through their rhizosphere. In addition, the phytotoxicity or volatilization of toxic chemicals through the leaves causes additional environmental problems. However, several plants grown in polluted environments host different types of associated bacteria able to degrade toxic organic pollutants. As plants can take up organic pollutants from contaminated media, through their root transport into their shoots and leaves, endophytic bacteria seem to be the best potential candidates for their degradation *in planta*. Endophytic bacteria are able to utilize several organic compounds as the sole carbon source for their growth and metabolism. The fate of organic pollutants in the root–rhizosphere system largely depends on their physicochemical properties. The octanol–water partition coefficient (*Kow*) is an important physicochemical characteristic widely used to describe hydrophobic and/or hydrophilic properties of organic pollutants. It is related to the transfer free energy of a compound from water to octanol. In the case of constant plant and environmental features, the *Kow* was shown to be the determining factor for root entry and translocation. Organic pollutants with a log *Kow* < 1 are considered very water soluble, and plant roots do not generally accumulate them at a rate surpassing passive influx into the transpiration stream, while organic pollutants with a log *Kow* > 3.5 show high sorption to the roots but slow or no translocation to the stems and leaves. However, plants readily take up organic contaminants with a log *Kow* between 0.5 and 3.5, as well as weak electrolytes such as weak acids and bases or

TABLE 4.1

Examples of Successful Remediation of Heavy Metal(s) by Using Plant–Endophyte Partnerships

Bacteria Species	Host Plant Species	Pollutants	Reference
41 Root endophytes, mainly affiliated with the phyla Proteobacteria, Bacteroidetes, Firmicutes, and Actinobacteria	*Betula celtiberica*	As	Mesa et al. (2017)
Bacillus pumilus E2S2, *Bacillus* sp. E1S2	*Sedum plumbizincicola*	Cd	Ma et al. (2015)
Bacillus thuringiensis GDB-1	*Alnus firma*	As	Babu et al. (2013)
Bacillus sp. SLS18	*Solanum nigrum*	Cd	Luo et al. (2011)
Burkholderia cenocepacia strain YG-3	*Populus deltoids* Marsh	Cd	Wang et al. (2019)
Burkholderia sp. HU001, *Pseudomonas* sp. HU002	*Salix alba*	Cd	Weyens et al. (2013)
Enterobacter aerogenes	*Brassica juncea*	Ni, Cr	Kumar et al. (2009)
Enterobacter ludwigii SAK5, *Exiguobacterium indicum* SA22	*Oryza sativa*	Cd, Ni	Jan et al. (2019)
Enterobacter sp. strain SVUB4	*Eichhornia crassipes*	Cd, Zn	El-Deeb et al. (2012)
Flavobacterium sp.	*Orychophragmus violaceus*	Zn	He et al. (2010b)
Kocuria sp. (LC2, LC3, and LC5), *Enterobacter* sp. (LC1, LC4, and LC6), *Kosakonia* sp. (LC7)	*Solanum nigrum*	As	Mukherjee et al. (2018)
Methylobacterium extorquens, Methylobacterium mesophilicum, Sphingomonas sp.	*Thlaspi goesingense*	Ni	Idris et al. (2004)
Methylobacterium oryzae CBMB20	*Lycopersicon esculentum* L.	Cd	Madhaiyan et al. (2007)
Micrococcus yunnanensis SMJ12, *Vibrio sagamiensis* SMJ18, *Salinicola peritrichatus* SMJ30	*Spartina maritima*	–	Mesa et al. (2015)
Neotyphodium coenophialum	*Festuca arundinacea, Festuca pratensis*	Cd	Soleimani et al. (2010)
Paenibacillus sp. SB12, *Bacillus* sp. SB31, *Bacillus* sp. LB51, *Alcaligenes* sp. RB54	*Chromolaena odorata*	Cd	Siripan et al. (2018)
Pantoea agglomerans Jp3-3, *Pseudomonas thivervalensis* Y1-3-9	*Brassica napus*	Cu	Zhang et al. (2011)

(Continued)

TABLE 4.1 (Continued)

Examples of Successful Remediation of Heavy Metal(s) by Using Plant–Endophyte Partnerships

Bacteria Species	Host Plant Species	Pollutants	Reference
Pseudomonas fluorescens G10, Microbacterium sp. G16	Brassica napus	Pb, Cd	Sheng et al. (2008a)
Pseudomonas putida PD1	Salix alba	Cd	Khan et al. (2014a,b)
Pseudomonas sp. M6, Pseudomonas jessenii M15	Ricinus communis	Ni, Cu, Zn	Rajkumar and Freitas (2008)
Rahnella sp. JN27	Amaranthus hypochondriacus, A. mangostanus, S. nigrum	Cd	Yuan et al. (2014)
Rhizobium leguminosarum	Brassica juncea	Zn	Adediran et al. (2015)
Serratia sp. RSC-14	Solanum nigrum	Cd	Khan et al. (2015)
Sphingomonas SaMR12	Sedum alfredii	Zn	Chen et al. (2014)
Staphylococcus saprophyticus, Massilia sp., Ochrobactrum intermedium, Bacillus sp., Bacillus pumilus, Staphylococcus sp., Aerococcus sp., Staphylococcus epidermidis, Pantoea stewartii, Ochrobactrum sp., Bacillus aerophilus, Microbacterium arborescens, Aerococcus viridans, Brevundimonas vesicularis, Enterobacter sp., Bacillus aquimaris, Pseudomonas sp.	Prosopis juliflora	Cr	Khan et al. (2014b)
Bacillus edaphicus NBT	Indian mustard	Pb	Sheng et al. (2008d)
Pseudomonas libanensis, Pseudomonas reactans	Brassica oxyrrhina	Cu, Zn	Ma et al. (2016b)

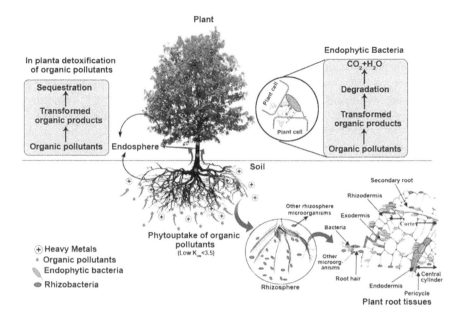

FIGURE 4.1 Mechanism of plants and their associated endophytic bacterial communities in degradation/detoxification of environmental pollutants.

amphoterics such as herbicides. These compounds seem to enter the xylem faster than the soil, and rhizosphere microflora can degrade them, even if the latter is enriched with degradative bacteria (Figure 4.1). Once these contaminants are taken up, plants metabolize these pollutants, although some of them or their metabolites can be toxic.

To avoid the toxicity associated with absorbed pollutants, plants usually follow one of two procedures, evapotranspiration or phytodegradation. For most pollutants, evapotranspiration is the major mechanism in which plants release pollutants in the atmosphere through their leaves. Endophytic bacteria have been shown to contribute to biodegradation of toxic organic compounds, and the plant–endophyte partnership can be exploited for the remediation of contaminated soil and groundwater (Weyens et al. 2009c,d). They are equipped with pollutant degradation pathways, and metabolic activities can diminish both phytotoxicity and evapotranspiration of volatile organic compounds (Weyens et al. 2009b,c; Shehzadi et al. 2015).

Endophytic bacteria, able to utilize several organic compounds as sole carbon source, were also isolated from the internal tissues of tree plants planted in polluted soil. Weyens et al. (2009c) isolated endophytic bacteria from English oak *Quercus robur* and common ash *Fraxinus excelsior* plants grown on a trichloroethylene (TCE)-contaminated site and found that the majority of the bacterial community showed increased tolerance to TCE and enhanced TCE degradation capacity in some of the strains. Cultivable endophytic bacteria isolated from bulk root, stem, and leaf were genotypically characterized and identified by 16S rRNA gene sequencing. A study by Moore and colleagues described 121 endophytic isolates from hybrid cottonwood (*Populus trichocarpa* × *deltoides* cv. "Hazendans" and "Hoogvorst") at a benzene, toluene, ethylbenzene, and xylene (BTEX)–contaminated site

(Moore et al. 2006). Some of the isolates demonstrated tolerance to heavy metals, BTEX, and TCE. An interesting finding was that isolates from leaf, stem, and root displayed different tolerances, suggesting that different microbial communities exist in different compartments of the plant. Indeed, nine of the isolates from poplar root could grow on BTEX and were tolerant to TCE, while seven isolates from the stem and only one of the isolates from leaves had these abilities. Out of the 121 endophytes studied, 34 were identified as having characteristics that might make them useful to enhance phytoremediation. Table 4.2 summarizes the role of endophytic bacteria in remediation of organic pollutant–contaminated sites.

4.7.3 ENDOPHYTIC BACTERIA IN ASSISTED PHYTOREMEDIATION OF MIXED POLLUTANTS

Although there exists an obvious difference in phytoremediation potential whether organics or metals are the primary targets, at most contaminated sites, plants and their associated microorganisms will have to deal with mixed contamination. Remediation of these mixed waste sites is generally intricate. The occurrence of toxic metals potentially inhibits a broad range of microbial processes, including the degradation of organic pollutants. A very promising strategy to tackle the mixed waste situation is the use of endophytes that are capable of (i) degrading organic contaminants and (ii) dealing with, or in the ideal scenario accelerating the extraction of, toxic metals. A study investigated the potential of indigenous endophytic bacteria to improve the efficiency of the wetland helophyte *Juncus acutus* to deal with mixed pollution consisting of emerging organic contaminants and metals. The beneficial effect of bioaugmentation with selected endophytic bacteria was more prominent in the case of high contamination: most of the inoculated plants (especially those inoculated with the mixed culture) removed higher percentages of organics and metals from the liquid phase in shorter times compared to the non-inoculated plants without exhibiting significant oxidative stress. The author indicates that the synergistic relationships between endophytes and the macrophyte enhance plants' performance and may be exploited in constructed wetlands treating water with mixed contaminations. It has been shown that engineering rhizobacteria for TCE degradation and Cd accumulation resulted in increased Cd accumulation but also in a lowered toxic effect of Cd on TCE degradation (Lee et al. 2006). Similar improvements are expected when these engineered rhizobacteria are inoculated onto plant roots.

4.8 ABUNDANCE, EXPRESSION, AND REGULATION OF CATABOLIC GENES FOR DEGRADATION OF ORGANIC POLLUTANTS

The population and diversity of pollutant-degrading endophytic bacteria usually depend on the presence of pollutant concentrations in soil and water. In addition to pollutant concentration, the population of pollutant-degrading endophytic bacteria also depends on the plant genotype, plant development stage, and environmental conditions. The pollutants degrading the ability of endophytes depend on the

TABLE 4.2

Examples of Successful Degradation of Organic Pollutants by Using Plant–Endophytic Bacteria Partnerships

Bacteria Species	Host Plant Species	Target Pollutants	Reference
Achromobacter xylosoxidans F3B	*Phragmites australis, Ipomoea aquatica*	Catechol, phenol	Ho et al. (2012)
Bacillus sp., *Microbacterium, Halomonas* sp.	*Typha domingensis, Pistia stratiotes, Eichhornia crassipes*	Textile effluent	Shehzadi et al. (2015)
Bacillus, Pseudomonas, Staphylococcus, Rhizobium, Lysinibacillus, Caulobacter, Paenibacillus, Aeromonas, Flavobacterium, Stenotrophomonas, Pantoea, Achromobacter, Sphingomonas, Rahnella	*Alopecurus aequalis, Oxalis corniculata* L.	PAHs	Peng et al. (2013)
Burkholderia cepacia	*Zea mays*	Phenol, toluene	Wang et al. (2010)
Burkholderia cepacia VM1468	Poplar	Toluene	Taghavi et al. (2005)
Burkholderia cepacia VM1468	*Lupinus luteus*	TCE	Weyens et al. (2010a)
Burkholderia cepacia VM1468	*Populus trichocarpa*	BTEX, TCE	Taghavi et al. (2011)
Burkholderia fungorum DBT1	Hybrid poplar (*Populus deltoides* × *Populus nigra*)	Dibenzothiophene, phenanthrene, naphthalene, fluorene	Andreolli et al. (2013)
Consortium CAP9	*Agrostis*	TNT	Thijs et al. (2014)
Enterobacter ludwigii strains	*Lolium multiflorum, Lotus corniculatus, Medicago sativa*	Diesel	Yousaf et al. (2011)
Enterobacter sp. 12J1	*Allium macrostemon*	Pyrene	Sheng et al. (2008c)
Methylobacterium populi	*Populus trichocarpa*	TNT, RDX, HMX	van Aken et al. (2011)
Methylobacterium populi BJ001	*Populus alba*	TNT, RDX, HMX	van Aken et al. (2004a,b)
Enterobacter sp. strain PDN3	*Populus trichocarpa*	TCE	Kang et al. (2012)
Pseudomonas, Acinetobacter, Microbacterium, Enterobacter, Pseudomonas, Rahnella	*Miscanthus giganteus, Iris pseudacorus*	BWW	Dunne et al. (2015)
Pseudomonas putida	*Pisum sativum*	Naphthalene	Germaine et al. (2009)

(Continued)

TABLE 4.2 (Continued)

Examples of Successful Remediation of Heavy Metal(s) by Using Plant–Endophyte Partnerships

Bacteria Species	Host Plant Species	Target Pollutants	Reference
Pseudomonas strains	Hybrid cottonwood (*Populus trichocarpa × Populus deltoides* cv. Hoogvorst)	2,4-D	Germaine et al. (2004)
Pseudomonas putida PD1	Willow and grass	Phenanthrene	Khan et al. (2014a)
Pseudomonas putida strain POPHV6/VM1450	*Pisum sativum*	2,4-D	Germaine et al. (2006)
Pseudomonas putida W619-TCE	*Populus alba*	TCE	Weyens et al. (2010b)
Pseudomonas sp.	*Lolium multiflorum*	Alkanes	Andria et al. (2009)
Pseudomonas sp. strain ITRI15, *Pseudomonas putida* W619-TCE	*Lolium multiflorum* var. Taurus, *Loltus corniculatus* var. Leo	TCE degradation	Yousaf et al. (2010)
Pseudomonas sp. strain ITRI53, *Rhodococcus* sp. strain ITRH43	*Lolium perenne*	Hydrocarbons	Andria et al. (2009)
Pseudomonas sp. strain ITRI53, *Pseudomonas* sp. strain MixRI75	*Lolium multiflorum*	Diesel	Afzal et al. (2011, 2012)
Pseudomonas sp., *Brevundimonas, Pseudomonas rhodesiae*	*Medicago sativa, Puccinellia nuttalliana, Festuca altaica, Lolium perenne, Thinopyrum ponticum*	n-Hexadecane	Phillips et al. (2008)
Pseudomonas tolaasii, P. jessenii, P. rhodesiae, P. plecoglossicida, P. veronii, P. fulva, P. oryzihabitans, Acinetobacter lwoffii, A. nicotianae, Bacillus megaterium, Paenibacillus amylolyticus	*Populus* cv. Hazendans	BTEX, TCE	Moore et al. (2006)
Rhodococcus erythropolis ET54b, *Sphingomonas* sp. D4	*Cytisus striatus*	HCH	Becerra-Castro et al. (2013)

(Continued)

TABLE 4.2 (*Continued*)
Examples of Successful Remediation of Heavy Metal(s) by Using Plant–Endophyte Partnerships

Bacteria Species	Host Plant Species	Target Pollutants	Reference
Rhizobiumtropici strain PTD1	*Populus* sp. (*Populus trichocarpa* × *P. deltoides*)	RDX	Doty et al. (2005)
Triticum spp.	*Herbaspirillum* sp. *K1*	PCBs, TCP	Mannisto et al. (2001)
Burkholderia cepacia VM1468	*Pisum sativum*	Toluene	Weyens et al. (2011)
Staphylococcus sp. BJ06	*Alopecurus aequalis*	Pyrene	Sun et al. (2014)
Rhizobium, Pseudomonas, Stenotrophomonas, Rhodococcus	*Lotus corniculatus* L., *Oenothera biennis* L.	Hydrocarbons	Pawlik et al. (2017)
Bacillus pumilus 2A	*Chelidonium majus* L.	Diesel oil, waste engine oil hydrocarbons	Marchut-Mikolajczyk et al. (2018)
Stenotrophomonas sp EA1-17, *Flavobacterium* sp. EA2-30, *Pantoea* sp EA4-40, *Pseudomonas* sp. EA6-5	White sweet clover	Hydrocarbons	Mitter et al. (2019)

Abbreviations: PCBs, Polychlorinated biphenyls; TCP, Trichlorophenol; BTEX, Benzene, toluene, ethylbenzene, and xylene; TCE, Trichloroethylene; TNT, Trinitrotoluene; 2,4-D, 2,4-Dichlorophenoxyacetic acid; PAHs, Polycyclic aromatic hydrocarbons; TNT, 2,4,6-Trinitrotoluene; RDX, Royal Demolition Explosive, hexahydro-1,3, 5-trinitro-1,3,5-triazine; HMX, octahydro-1,3,5,7-tetranitro-1,3,5-tetrazocine; HCH, Hexachlorocyclohexane; BWW, Brewery wastewater.

expression of genes that encode particular enzymes that degrade and/or mineralize organic compounds. Sessitsch et al. (2012) found through metagenomic analysis that endophytic bacteria isolated from the roots of rice plants have the potential to degrade aliphatic and aromatic hydrocarbons. Additionally, genes encoding the production of plant polymer–degrading enzymes, quorum sensing, and detoxification pathways were found. This metagenomic analysis revealed that degradation genes were highly abundant, although rice plants and thus also endophytes were not exposed to a polluted environment. This may indicate that endophytes may be equipped with genes to degrade complex organic substances, which may be produced as plant metabolites and utilized as carbon sources by the bacteria. The same enzymes might be involved in the degradation of pollutants. Similarly, the genome survey of *Burkholderia phytofirmans* PsJN, a well-known plant growth-promoting endophytic bacterium, revealed that this strain also possesses hydrocarbon-degrading alkane monooxygenase (*alkB*) and alkane hydroxylase (*CYP450*) genes (Mitter et al. 2013). These studies have shown that many endophytic bacteria have multiple traits that are useful for the remediation of contaminated soils. To our knowledge, this strain has not been exposed to polluted environments, and degradation genes may contribute to the extremely high capacity to colonize plant tissues endophytically. The endophyte strain *Pseudomonas* sp. ITRI53 exhibited after inoculation the highest levels of abundance and expression of *alkB* genes in the roots and shoots of ryegrass vegetated in loamy soil, whereas the lowest levels of gene abundance and expression were observed in the shoot and root of ryegrass vegetated in sandy soil. Moreover, this endophytic bacterial strain exhibited metabolic activity in the endosphere of ryegrass planted in the sandy soil but not in the rhizosphere of this grass. In another recent study, three different plant species (Italian ryegrass, birdsfoot trefoil, and alfalfa) were inoculated with three different endophytic *Enterobacter ludwigii* strains (Yousaf et al. 2011). Generally, the inoculated strains showed higher levels of colonization, abundance, and expression of alkane-degrading *CYP153* genes in shoots and roots of Italian ryegrass than alfalfa and birdsfoot trefoil. Moreover, the abundance and expression of *CYP153* genes of all inoculated strains varied markedly between different strains, plant species, plant compartments, and plant development stages. Different plants host different endophytic bacteria to a different extent and also stimulate their pollutant-degradation activity differently. It has been also demonstrated that the inoculation method may affect endophytic bacterial colonization and activity during the phytoremediation of diesel-contaminated soil. The endophyte niche is a hot spot for horizontal gene transfer (HGT), as demonstrated by Taghavi et al. (2005). In this study, the degradative plasmid pTOM-Bu61 was found to have transferred naturally to a number of different endophytes *in planta*. This HGT activity promoted more efficient degradation of toluene in poplar plants. HGT *in planta* is likely to be widespread, as studies in pea with *Pseudomonas* endophytes harboring the plasmids pWWO and pNAH7 also show high rates of transfer into a range of autochthonous endophytes. This approach may have practical applications in equipping natural endophyte populations with the capacity to degrade a pollutant and does not require long-term establishment of the inoculant strain. It has been demonstrated that endophytic bacteria efficiently expressing the necessary catabolic genes can promote the degradation of xenobiotic compounds or their metabolites as they are accumulated or while being translocated in the vascular tissues of the host plant.

5 Diazotroph-Assisted Phytoremediation of Heavy Metals

5.1 INTRODUCTION

In recent years, heavy metal (HM) contamination has become a serious concern to the environment and living organisms, that is, plants, animals, and microorganisms, as HMs persist in the environment for longer periods due to their nonbiodegradability (Chandra et al. 2018a). Although a broad range of physico-chemical methods have been applied for remediation of soil and water contaminated with heavy metals, these methods are costly and not environmentally safe. Phytoremediation is an emerging cost-effective technology for remediating sites contaminated with toxic HMs. This technology has been considered the most promising due to its minimal site disturbance, low cost, and higher public acceptance when compared with conventional remediation methods (Cang et al. 2011; Meier et al. 2012; Luo et al. 2014). Phytoremediation employs plants, alone or together with their associated microorganisms, to degrade, contain, or stabilize various HMs in the environment. Therefore, phytoremediation, together with the use of microbes, facilitates greater remediation of HMs as compared to phytoremediation alone (Weyens et al. 2009a; Rajkumar et al. 2012; Nanekar et al. 2015; Thijs et al. 2017). However, the outcome of phytoremediation technology depends on microorganisms (type of microorganisms, population diversity, and their enzyme activities), exudates (microbial and plant), soil (physico-chemical characteristics, structure, and concentration), and a range of environmental factors (pH, temperature, moisture content, availability of electron acceptors, and carbon and energy sources) (Jing et al. 2007; Phillips et al. 2008; Khan et al. 2014a,b; Yang et al. 2016; Ahemad 2019). All these factors affect the phytoremediation of HMs. Plant-associated bacteria (rhizobacteria and endophytic bacteria) help in the minimization of these phytoremediation factors, as they have plant growth-promoting characteristics that assist phytoremediation (Ullah et al. 2015a,b). Bacteria capable of assimilating and fixing atmospheric dinitrogen (N_2) into a biologically useable ammonium form (NH_4^+) are referred to as diazotrophs (Weyens et al. 2009d, 2012; Wagner 2011). In addition, diazotrophs improve plant growth through phytohormone production, specific enzymatic activity, and plant protection from diseases by the production of antibiotics or other pathogen-suppressing substances such as siderophores and antibiotics. Some diazotrophs also have the ability to decrease the level of ethylene in plants through the production of aminocyclopropane-1-carboxylic acid (ACC) deaminase, which ameliorates the operation of phytoremediation (Wagner 2011; Ullah et al. 2014, 2015b). The aim of the present chapter is to provide a concise discussion of the role and mechanisms of diazotrophs in assisting phytoremediation of HMs in contaminated environments.

5.2 DIAZOTROPHS, DIAZOTROPHIC BACTERIA, AND DIAZOTROPHY

Nitrogen (N) is the most abundant and vital element in the Earth's atmosphere (nearly 78% in form of diatomic molecules [N_2 or $N = N$]) and essential for plant growth and metabolism. It is a major component of chlorophyll, the most important pigment needed for photosynthesis, as well as amino acids (Wagner 2011). It is also a critical limiting element available for the growth of producers and consumers in all ecosystems because gaseous N_2 is chemically inert and cannot be assimilated and thus needs to be converted into soluble and assimilated forms like ammonium (NH_4^+), nitrite (NO_2^-), or nitrate (NO_3^-). Among them, NH_3 is the most assimilated form of N by plants. It is not always bioavailable in soil, as most N is present as N_2 in soil porosity or as part of humic compounds (Wagner 2011; Temmink et al. 2018). Among prokaryotes, diazotrophs can facilitate plant N nutrition through biological N fixation (BNF), which corresponds to the reduction of atmospheric N_2 to NH_3. BNF, the ability to reduce atmospheric N_2 gas to NH_3, discovered by Beijerinck in 1901 (Beijerinck 1901), is carried out by a specialized group of prokaryotic microorganisms (bacteria, cyanobacteria, archaea) collectively known as diazotrophs. It is a major source of fixed N to the biosphere performed by a diverse array of diazotrophs: they provide about 60% of the total annual input. The great diversity of diazotrophs also extends to their physiological characteristics, as N fixation is performed by chemotrophs and phototrophs and by autotrophs as well as heterotrophs (Wagner 2011; Norman and Friesen 2017). Diazotrophic bacteria are generally grouped on the basis of their lifestyle, being either free living, symbiotic, or in associative symbiosis, usually with plant roots (free-living bacteria include *Azospirillum, Bacillus, Azotobacter, Beijerinckia, Pseudomonas, Klebsiella, Bacillus,* and *Clostridium*) (Figure 5.1). In addition, the species of photosynthetic bacteria *Rhodobacter, Rhodospirillum, Thiospirillum,* and *Chromatium*; different species of methanogenic archaebacteria such as *Methanococcus, Methanosarcina, Methanobacter,* and *Methanospirillum*; and *Cyanobacteria* species of *Anabaena, Nostoc,* and *Calothrix* are also known to be

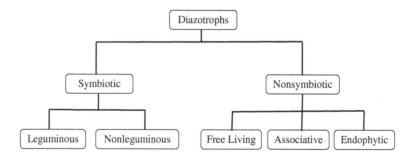

FIGURE 5.1   Categorization of nitrogen-fixing microorganisms.

free-living potential nitrogen-fixing diazotrophs (Wagner 2011; Ullah et al. 2014). It is generally believed that free-living bacteria provide only a small amount of the fixed N that the bacterially associated host plant requires (Burgmann et al. 2004). Because nitrogenase can be inhibited by oxygen, free-living organisms behave as anaerobes or microaerophiles while fixing N_2. *Klebsiella pneumoniae* was the first diazotroph in which the N_2 fixation process was analyzed at the molecular level, primarily because most of the genetic tools developed for *Escherichia coli* could be used in this closely related enterobacterium (Fouts et al. 2008). Species of *Azospirillum* are able to form close associative relationships with several members of the Poaceae, including agronomically important cereal crops, such as rice, wheat, corn, oats, and barley, and fix significant amounts of N_2 within the rhizosphere of the host plants (Steenhoudt and Vanderleyden 2000). Facultative and obligate endophytic diazotrophs colonizing non-leguminous plants include several bacteria belonging to the genus, namely *Azospirillum* (*A. brasilense, A. lipoferum, A. amazonense, A. irakense*) and the endophytes *H. seropedicae, H. rubrisubalbicans, A. diazotrophicus, Azoarcus* spp., and *Burkholderia* spp. isolated from sugarcane, palm trees, forage grasses, tuber plants, cereals, and sweet potato fixed N_2 (Steenhoudt and Vanderleyden 2000). Many microorganisms fix N_2 symbiotically by partnering with a host plant. One example of this type of N_2 fixation is the water fern *Azolla*'s symbiosis with the cyanobacterium *Anabaena azollae* (Naghipour et al. 2018). The diazotrophs live inside *Azolla's* leaf cavities and include the cyanobacteria *Nostoc/Anabaena azollae* that form unbranched, multicellular chains that contain both photosynthetic vegetative cells and N_2-fixing heterocysts. Actinorhizal species (e.g., *Alnus, Casuarina, Myrica*) form symbiotic N_2-fixing nodules (rhizothamnia) with *Actinobacteria*, and cycads (e.g., *Ceratozamia, Macrozamia*) form N_2-fixing structures (coralloid roots) with cyanobacteria. Using nitrogenase enzymes, the diazotrophs reduce atmospheric N_2 to NH_4^+, which is then excreted into the *Azolla* leaf cavity and taken up by the fern. Even though the symbiotic partners described previously play an important role in the worldwide ecology of N_2 fixation, by far the most important N_2-fixing symbiotic associations are the relationships between legumes and bacterial members of the family Rhizobiaceae (Wagner 2011). Members of the family Rhizobiaceae are able to form symbiosis with leguminous plants like rhizobia and non-leguminous trees such as Frankia. The legume–*Rhizobium* symbiosis under stressful conditions not only facilitates the growth of legume plants but also increases soil fertility. In recent years, a great diversity of highly heavy metal–resistant rhizobial strains, including *Azorhizobium, Bradyrhizobium, Mesorhizobium, Sinorhizobium*, and *Rhizobium*, have been isolated from industrial areas (Gage 2004). These Rhizobia have the capability to infect roots and induce the formation of root nodules in plants from family Leguminosae: alfalfa, beans, clover, cowpeas, lupines, peanut, soybean, and vetches. Root nodules are the sites for nitrogen fixation formed on plant roots (Wang et al. 2018). This interaction consists of several stages that involve the exchange of complex signals between the bacterium and plant. To establish an N_2-fixing symbiosis, rhizobia respond to plant-derived flavonoids in the rhizosphere by synthesizing Nod factor, a lipochitooligosaccharide signal molecule that elicits both nodule organogenesis and root hair deformation in the plant. Nod factors are perceived by plant root hairs and function in a hormone-like fashion to induce root nodules in which the *Rhizobium* bacterium can fix atmospheric N_2. The bacterium grows at the expense of carbohydrates from the host but provides fixed N_2 for amino acid biosynthesis in return. The products of these genes (Nod factors) induce root-hair curling on the plant and cortical cell divisions, which are among the earliest microscopically observable events in the nodulation of most legume species (Geurts and Bisseling 2002). When rhizobia adhere to root hairs, the cell wall of the affected root hair is partly hydrolyzed at the tip, allowing the bacteria to enter. An infection thread is formed by invagination of the cell wall. Alternatively, rhizobia may enter through cracks in the epidermis, associated with lateral-root formation, or wounds (Wang et al. 2018). However, in that event, effective signaling must also occur for a nodule to emerge. Diazotrophs catalyze N_2 fixation via a complex enzyme system called nitrogenase, which has been highly conserved throughout evolution. This nitrogenase enzyme system exists as two metalloprotein components: (i) dinitrogenase reductase (Fe-protein) and (ii) nitrogenase, or nitrogenase molybdenum-iron (MoFe) protein (Norman and Friesen 2017; Rodrigues et al. 2017). The former enzyme serves as an exclusive electron donor with high reducing power, whereas the latter (substrate reduction component) accepts the electron's energy and converts inert N_2 molecule into ammonia (NH_3). In order to produce 1 mole of NH_3, 16 moles of adenosine triphosphate (ATP) are required by these diazotrophs (Hubbell and Kidder 2009), which obtain this energy by oxidizing organic molecules.

$$N_2 + 16ATP + 8e^- + 8H^+ \rightarrow 2NH_3 + H_2 + 16ADP + 16Pi$$

In diazotrophic bacteria, *nif* (nitrogen fixation) genes, namely *nif*A, *nif*B, *nif*D, *nif*E, *nif*F, *nif*H, *nif*K, *nif*V, *nif*M, and so on, which encode the nitrogenase enzyme complex, are typically found in a cluster of around 20–24 kb with seven operons encoding 20 different proteins (Wagner 2011; Bouffaud et al. 2016; Norman and Friesen 2017). Nitrogenase is an oxygen-sensitive enzyme, and its activity is inhibited in presence of oxygen; this means that N_2 fixation takes place under anoxic conditions (Burgmann et al. 2004). Among the numerous *nif* genes, the *nif*H gene encoding the iron protein subunit of nitrogenase is highly conserved among diazotrophs, and its phylogeny is largely correlated to 16S rRNA phylogeny, making it a marker suitable to assess complex diazotrophic communities in various environments (Burgmann et al. 2004; Norman and Friesen 2017). In ecosystems, diazotrophic community size and/or composition can vary according to soil type, soil management, season, fertilization rate, plant presence/absence, plant growth stage, plant species, or plant varieties (Rodrigues et al. 2017; Ma et al. 2019). Some of these factors may also influence the composition of the bacterial community actually expressing *nif*H and/ or the amount of *nif*H transcripts (Burgmann et al. 2004)

5.3 ROLE OF PLANT–MICROBE INTERACTIONS IN PHYTOREMEDIATION AND NITROGEN FIXATION

The interaction between bacteria and plants occurs in different niches, such as in soil near the root surface (rhizospheric community) and within plants (endophytic community) (Hartmann et al. 2009; Shin et al. 2016; Rodrigues et al. 2017). Bacteria can enhance the growth of different organs such as the root, stem, leaf, and flowers and contribute to plant-growth promotion and development even in metal-contaminated soils (Al-Awadhi et al. 2009; Gullap et al. 2014). The bacteria that promote plant growth are known as plant growth-promoting bacteria (PGPB) (Glick 2012). Plant growth-promoting rhizobacteria (PGPR) and plant growth-promoting endophytic bacteria (PGPE) are the two main types of soil bacteria that have been shown to act as PGPB (Wu et al. 2006b; Lugtenberg and Kamilova 2009; Ullah et al. 2014). During beneficial interactions, endophytic and rhizospheric bacteria can exert several functions, such as BNF; improved ammonia production; synthesis of siderophores as iron chelators that can solubilize and sequester iron from the soil and provide it to plants—it has been suggested that these compounds play a role in production of phytohormones such as indole acetic acid (IAA), gibberellic acid, or cytokinins that increase root growth; production of aminocyclopropane-1-carboxylic acid deaminase, which hydrolyzes ACC, the biosynthetic precursor for ethylene in plants, into ammonia and α-ketobutyrate to reduce the level of ethylene in the roots of developing plants; production of quorum sensing (QS) molecules; antifungal metabolites or lytic enzymes; production of exopolysaccharides and osmoprotectants; and solubilization of inorganic phosphate and mineralization of organic phosphate and/or other nutrients, which then become more readily available for plant growth (Xin et al. 2009; Knoth et al. 2013; Ma et al. 2016a,b; Shin et al. 2016). Interaction of some PGPB with plant roots can result in plant resistance against some pathogenic bacteria, fungi, nematodes, and viruses. This phenomenon is called induced systemic resistance (ISR). All these interactions can contribute to plant-growth promotion. Free-living N_2-fixing bacteria were probably the first bacteria used to promote plant growth. *Azospirillum* was first isolated in the 1970s (Steenhoudt and Vanderleyden 2000), and this genus has been studied widely, the study by Bashan et al. (2004) being the most recent one. It reported the latest advances in physiology, molecular characteristics, and agricultural applications of this genus. Other bacterial genera capable of the N_2 fixation that is probably responsible for the plant-growth promotion effect are *Azoarcus* sp., *Burkholderia* sp., *Gluconacetobacter diazotrophicus, Azotobacter* sp., *Herbaspirillum* sp., and *Paenibacillus polymyxa*. Plant roots are colonized by a variety of bacterial species from different genera, such as *Bacillus, Paenibacillus, Burkholderia, Azotobacter, Rhizobium,* and *Pseudomonas*, which simultaneously function together to synergistically promote plant growth (Antoun and Kloepper 2001; Glick 2012; Ahemad 2019). PGPR isolated as free-living soil bacteria from the plant rhizosphere can decrease chemical fertilizer-N use and increase plant growth and yield when associated with plant roots and other plant parts (Beneduzi et al. 2012; Glick 2012; Bashan et al. 2013; Kong and Glick 2017). Several bacteria, such as *Azospirillum, Klebsiella, Burkholderia, Bacillus,* and *Pseudomonas*, have been identified as PGPR to maize plants through BNF, phosphate solubilization, phytohormone production (e.g., auxin, gibberellin, and cytokinin), and biological control of soil pathogens. Metal-resistant PGPB have been widely investigated for their potential to alleviate metal toxicity and immobilize/mobilize/transform metals in soil, which may help to develop new diazotroph-assisted phytoremediation and restoration strategies (Ullah et al. 2014; Shin et al. 2016; Kong and Glick 2017).

In addition to N fixation, the plant growth-promoting abilities expressed by metal-resistant rhizobia affect metal solubility and bioavailability, both of which affect plant metal uptake. Some soil bacteria may facilitate the adaptation of host plants to suboptimal soil conditions and enhance the efficiency of phytoremediation by promoting plant growth, alleviating metal phytotoxicity, altering metal bioavailability in soil, and increasing metal translocation within the plant (Shin et al. 2016). Also, PGPB may alter metal accumulation capacity and its translocation in plants by their multiple plant growth-promoting traits, including metal resistance, detoxification, accumulation, transformation, and sequestration, thus diminishing the metal phytotoxicity and altering the phytoavailability of HMs in contaminated soils (Ma et al. 2016a).

It is responsible for the nutrient dynamics in soil plant systems that affects the soil-to-soil solution and also solution to plant-root movement/availability. The occupation of metals in native binding sites of microbial cells that are specifically for essential nutrients or metals and through ligand interactions results in metal toxicity in microbes. For instance, Hg^{2+}, Cd^{2+}, and Ag^{2+} are likely to bind with sulfhydryl (–SH) groups of some sensitive enzymes and hinder the function of the enzymes. But at higher concentrations, whether essential or nonessential, metal can damage the membranes of the microbial cell wall and interrupt the function of the cells by damaging DNA structure and altering enzyme specificity. However, some HM-resistant microbes are adaptive to HM-rich environments. Volatile methylated species are often lost from the soil. Several bacteria can methylate methylmercury, forming volatile dimethylmercury. Methylmercury, as well as phenylmercury, can be enzymatically reduced to volatile metallic mercury, Hg^0, by some bacteria. Phenylmercury can also be microbially converted to diphenylmercury. Generally, bacterial cells uptake the heavy metal cations of similar size, structure, and valence with the same mechanism. Bacteria generally possess two types of uptake systems for heavy-metal ions: One is fast and unspecific and driven by the chemiosmotic gradient across the plasma membrane, and another type is slower, exhibits high substrate specificity, and is coupled with ATP hydrolysis. Microorganisms that are highly beneficial and play an important role in providing nutrients can reduce the noxious effects of metals on plants. Some of these rhizosphere microorganisms can work directly on organic and inorganic pollutants using their own degradation capabilities, for instance, volatilization, transformation, and rhizodegradation.

5.4 PHYTOREMEDIATION OF HEAVY METALS FACILITATED BY DIAZOTROPHS

Diazotrophs associated with plants can enhance the phytoremediation process directly by changing metal bioavailability, releasing chelating compounds (organic acids, siderophores) and methylation, altering soil pH, and redox reactions (Liu et al. 2015). Such microbes improve HM mobilization and solubility, which further increase the uptake of HMs by different biogeochemical processes (Ullah et al. 2014; Shin

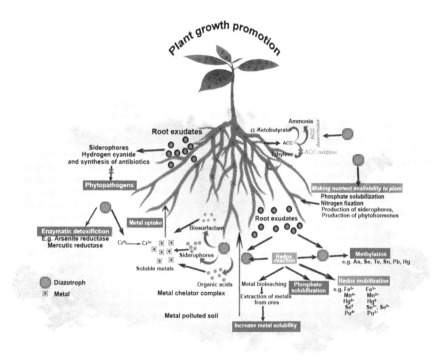

FIGURE 5.2 An overview of the mechanism used by diazotrophs to mobilize nutrients in the soil.

et al. 2016). These biogeochemical cycles include processes such as transformation, translocation, chelation, immobilization, solubilization, volatilization, precipitation, and the complexation of HMs, finally facilitating phytoremediation (Ullah et al. 2014). These biogeochemical cycles are completed with the help of different secretions from bacteria such as siderophores, organic acids, biosurfactants, polymeric substances, and glycoprotein. PGPB are the scavengers of contaminants and reduce metal toxicity in soil toward plants. To sum up, the mechanisms PGPB employ to promote the phytoremediation process involve improvement of plant metal tolerance and increased plant growth, as well as alteration of metal accumulation in plants (Figure 5.2 and Table 5.1). The mechanisms of plant-growth promotion have been discussed previously. For phytoremediation of toxic metals, endophytes possessing a metal-resistance/sequestration system can lower metal phytotoxicity and affect metal translocation to above-ground plant parts (Weyens et al. 2009a,b; Abhilash et al. 2012; Huo et al. 2012; Yuan et al. 2014). PGPR can enhance plant growth through nutrient recycling; nitrogen fixation; phytohormone production; solubilization of nutrients such as P, K, and Fe; and enhancing plant resistance to pests and diseases (Glick 2014; Gouda et al. 2018; Ahemad 2019). Therefore, the primary aim of PGPR addition is to increase overall yield. Diazotrophic bacteria, however, can fix atmospheric nitrogen, even under metal-stressed conditions (Shin et al. 2016).

TABLE 5.1

Selected References on Phytoremediation of Metals Facilitated by Soil Bacteria

Bacterial Strain(s)	Host Plant Species	Pollutants	Source of Isolation	Beneficial Effects	Reference
Achromobacter xylosoxidans Ax10	Brassica juncea	Cu	–	Increased root and shoot length and biomass; ACCD, Psolubilization, IAA	Ma et al. (2009b)
Acinetobacter sp. NBRI05	Cicer arietinum	As^{5+}, As^{3+}	As-contaminated soil	ACCD activity, IAA, Psolubilization, siderophores	Srivastava and Singh (2014)
Acinetobacter sp. RG30, Pseudomonas putida GN04	Zea mays	Cu	Heavy metal– and hydrocarbon-contaminated soil	IAA, Psolubilization, siderophores	Rojas-Tapias et al. (2014)
Azotobacter chroococcum HKN-5 + Bacillus megaterium HKP-1 + Bacillus mucilaginosus HKK-1	Brassica juncea	Zn, Cu, Pb, Cd	–	Increased biomass and metal bioavailability	Wu et al. (2006a)
Bacillus licheniformis, Bacillus biosubtyl, Bacillus thuringiensis	Brassica juncea L.	Se, Cd, Cr	Contaminated environment	Increased metal uptake, mechanism unknown	Hussein (2008)
Bacillus pumilus E2S2	Sedum plumbizincicola	Cd, Zn, Pb	Zn/Cd hyperaccumulator Sedum plumbizincicola	ACCD activity, P solubilization, IAA, siderophores	Ma et al. (2015)
Bacillus sp. SLS18	Sorghum bicolour L., Phytolacca acinosa Roxb, Solanum nigrum L.	Mn, Cd	Stem of Mn hyperaccumulator Phytolacca acinosa	IAA, siderophores, ACCD	Luo et al. (2012)
Burkholderia sp. RX232	Salix caprea	Zn, Cd, Pb	Rhizosphere soil of Salix caprea	ACCD activity, siderophores	Kuffner et al. (2010)

(Continued)

TABLE 5.1 (Continued)

Selected References on Phytoremediation of Metals Facilitated by Soil Bacteria

Bacterial Strain(s)	Host Plant Species	Pollutants	Source of Isolation	Beneficial Effects	Reference
Cu-resistant isolates belonging to Firmicutes, Actinobacteria, and Proteobacteria	Brassica napus	Cu	tissues of Elsholtzia splendens and Commelina communis	IAA, siderophores, ACCD, and arginine decarboxylase	Sun et al. (2010)
Enterobacter aerogenes NBRI K24, Rahnella aquatilis NBRI K3	Brassica juncea	Ni, Cr	Fly ash contaminated soil	Increased biomass and metal uptake, IAA, siderophores, ACCD activity, Psolubilization	Kumar et al. (2009)
Enterobacter sp. NBRI K28	Brassica juncea	Ni, Zn, Cr	Fly ash contaminated soil	Increased biomass and metal uptake; IAA, siderophores, ACCD, Psolubilization	Kumar et al. (2008)
Klebsiella ascorbata SUD165, SUD165/26	Brassica napus, Lycopersicon esculentum, Brassica juncea	Ni, Pb, Zn	Contaminated soil	Increased biomass; ACCD, siderophores	Burd et al. (2000)
Kluyvera ascorbata SUD165	Brassica napus	Ni	Contaminated soil	Increased biomass; ACCD	Burd et al. (1998)
Microbacterium sp. EX72	Salix caprea	Zn, Cd	Tissues of Salix caprea	Increased concentrations of Zn and Cd in leaves	Kuffner et al. (2010)
Microbacterium arabinogalactanolyticum	Alyssum murale	Ni	Rhizopsphere of A. murale	Increased nickel uptake; mechanism unknown	Abou-Shanab et al. (2003)
Pseudomonas putida 06909	Helianthus annuus, Vigna unguiculata, Triticum sativum, Zea spp	Cd	Citrus root	Increased Cd uptake and decreased toxicity; bacterium expresses a metal-binding peptide	Wu et al. (2006a)
Pseudomonas sp. TLC 6-6.5-4	Zea mays and Helianthus annuus	Cu	Torch lake sediment	IAA, Pand metal solubilization	Li and Ramakrishna (2011)

(Continued)

TABLE 5.1 (Continued)
Selected References on Phytoremediation of Metals Facilitated by Soil Bacteria

Bacterial Strain(s)	Host Plant Species	Pollutants	Source of Isolation	Beneficial Effects	Reference
Pseudomonas sp. A4, *Bacillus* sp. 32	*Brassica juncea*	Cr	Heavy metal contaminated soil	Increased root and shoot length; IAA, siderophores, P solubilization	Rajkumar et al. (2006)
Pseudomonas sp. M6, *Pseudomonas jessenii* M15	*Ricinus communis*	Ni, Cu, Zn	–	Increased biomass; IAA, ACCD production, phosphate solubilization	Rajkumar and Freitas (2008)
Pseudomonas asplenii AC	*Brassica napus*	Cu	PAH contaminated soil	Increased biomass; IAA	Reed and Glick (2005)
Rahnella sp. JN27	*Solanum nigrum, Zea mays, Amaranthus hypochondriacus, Amaranthus mangostanus*	Cd	Roots of *Zea mays*	IAA, siderophores, ACCD production, P solubilization	Yuan et al. (2014)
Ralstonia eutropha, Chryseobacterium humi	*Zea mays* L.	Cd	Sediment from an industrially contaminated site	–	Moreira et al. (2014)
Rhizobium sp. RL9	*Lens culinaris* var. Malka	Cu, Cd, Cr, Ni, Pb, Zn	Root nodules of lentil plants	IAA, siderophores, HCN, ammonia	Wani and Khan (2013)
Rhodococcus erythropolis NSX2	*Sedum plumbizincicola*	Cd, Cu	Roots of *Sedum X Graptosedum*	IAA, P solubilization	Liu et al. (2015)

Abbreviations: HCN, Hydrogen cyanide; IAA, Indole acetic acid; P, Phosphate; ACCD, Aminocyclopropane-1-carboxylic acid deaminase; PAH, Polycyclic aromatic hydrocarbons.

6 Phosphate-Solubilizing Microbe-Assisted Phytoremediation of Heavy Metals

6.1 INTRODUCTION

Soil and water pollution by toxic heavy metals (HMs) has been accelerated greatly by the use of HMs in various industrial activities such as distilleries, pulp-paper manufacture, tanning, mining, electroplating, paint, cement, ceramic refining, and manufacturing processes (Wuana and Okieimen 2011; Kumar and Chandra 2020a). Also, HMs are also used in different fungicides and land chemical fertilizers, wastewater irrigation, and sewage sludge, causing HM contamination of agricultural soils and water resources (Chandra et al. 2018a,b). Massive amounts of HMs discharged from various sources are continually entering food chains and consequently severely affecting metabolisms, leading eventually to the death of microbes, plants, and animals (Tchounwou et al. 2012). Thus, remediation of HMs is necessary to protect the environment from their toxic effects. Phytoremediation is considered a novel environmentally friendly technology, which uses plants to remove or immobilize HMs from contaminated sites (Salt et al. 1995; Cabral et al. 2015). This technique is based on the use of plants to clean up and/or improve soil and water quality by inactivation or translocation of pollutants in different parts of the plant, without negative effects on the structure, fertility, or biological activity of the soil. However, this process suffers from some limitations, such as the low mobility of the tightly bound fraction of heavy metals in soils with neutral pH, resulting in reduced uptake by plants (Rajkumar et al. 2012; Ojuederie and Babalola 2017). In addition, some HMs are toxic to plants even at very low concentrations. To address such limitations, scientists have developed microbe-assisted phytoextraction, in which bacteria are added to soil to facilitate HM phytoextraction. Interestingly, interactions between plants and metal-resistant microbes, especially bacteria, have shown better remediation of HMs, and this synergism not only expedites the remediation process by ameliorating phytostabilization and phytoextraction of metal species but also accelerates plant growth and development in HM stress environments (Tirry et al. 2018; Ahemad 2019). It has been reported that several phosphate-solubilizing microbes (PSMs) exhibiting both heavy metal–detoxifying traits and plant growth-promoting activities can improve plant growth in metalliferous soils by supplying phosphorus, and they are thus beneficial to phytoremediation of metalliferous soils (Rodríguez and Fraga 1999).

This chapter is an effort to emphasize how the beneficial association between plants and PSMs can be used to remediate metal-stressed soils efficiently. In this chapter, the mechanism of PSM mediation in supporting and intensifying the phytoremediation process is discussed in detail.

6.2   PHOSPHATE-SOLUBILIZING MICROBES IN MANAGEMENT OF PHOSPHOROUS-DEFICIENT SOIL

Phosphorus (P) is the second most essential macro-mineral for the growth and development of plants, next to nitrogen (N); it makes up about 0.2% of a plant's dry weight. It plays an important role in virtually all major metabolic processes in plants, including photosynthesis, proper plant maturation, stress mitigation energy transfer, signal transduction respiration, and nitrogen fixation in legumes (Ahemad 2015). It is also an essential component of adenosine triphosphate (ATP), nucleic acids (DNA, RNA), and phospholipids. Despite the fact that the amount of P in the soil is relatively high (its concentration in soil often varies between 400 and 1,200 mg kg^{-1} of soil) (Chen et al. 2008), most of this P is insoluble and thereby not bioavailable to plant uptake: only a very small fraction ($\sim$0.1%) is available to plants to support plant growth. This insoluble P is present in the environment as either an inorganic mineral, namely apatite (the original source of all phosphorus) or as one of several organic forms, including phosphotriesters, phosphomonoesters, and inositol phosphate (soil phytate) (Bashan et al. 2013). In soil, P always forms complexes with other compounds in the form of phosphates. Inorganic forms of soil P (Pi) consist of apatite complexed with iron (Fe), calcium (Ca), aluminum (Al), phosphate, and P absorbed onto clay particles. The fixation of Pi into insoluble complexes renders these compounds inaccessible for absorption by plants and therefore results in a severe Pi insufficiency in both acidic and alkaline soils. Plants can only take up P in two soluble forms, that is, monobasic ($H_2PO_4^-$) or dibasic (HPO_4^{2-}) (Rodríguez and Fraga 1999; Zaidi et al. 2009). To address the problem of P deficiency in different crops, phosphatic fertilizers are added in various amounts to soil. These phosphatic fertilizers use rock phosphate as the main source of P_2O_5. In addition, much of the soluble inorganic P that is used as chemical fertilizer is immobilized soon after it is applied so that it then becomes unavailable to plants and is therefore wasted. Traditionally, the challenge of soil phosphorus deficiency is addressed by the application of phosphorus fertilizers (Sharma et al. 2013). However, the majority of the applied fertilizer P is not available to plants, and the addition of inorganic fertilizers in excess of the amount that is commonly employed to overcome this effect can lead to environmental problems such as groundwater contamination and waterway eutrophication (Kang et al. 2011). It is therefore of great interest to investigate management strategies that are capable of improving phosphorus fertilization efficiency, increasing crop yields, and reducing environmental pollution caused by P loss from the soil. Apart from chemical fertilization, microbial P-solubilization and mineralization are the only possible ways to increase plant-available P (Zaidi et al. 2009; Panhwar et al. 2013). These fertilizers also pose a serious environmental threat, and thus alternative strategies are being developed for sustainable agriculture (Alori et al. 2017). In recent years, the use of phosphate-solubilizing microorganisms in crop production, as an alternative method to meet crop phosphorus demand, has received increased attention. In this context, organisms endowed with P-solubilizing activity may provide a viable substitute to chemical P fertilizers (Kalayu 2019). The use of microbes as biofertilizers, on the other hand, has gained much interest in the recent era due to their promising effect on growth and yield of plants as well as soil fertility. A large number of microbes, including bacteria, fungi, actinomycetes, and algae, have exhibited P solubilization and mineralization ability (Zaidi et al. 2009). In addition to bacteria, fungi, and actinomycetes, algae such as cyanobacteria and mycorrhiza have also been reported to show P solubilization activity (Ahemad 2015; Alori et al. 2017).

6.2.1 Phosphate-Solubilizing Bacteria

Phosphate-solubilizing bacteria (PSB) are predominant microorganisms that play an important role in biogeochemical P cycling in both terrestrial and aquatic environments, as compared to other microorganisms. PSB can transform insoluble P to available P in the soil so as to improve fertilizer use efficiency and crop yield. PSB secrete organic acids that dissolve unavailable P (PO_4^{3-}) to plant-available forms such as HPO_4^{2-} and $H_2PO_4^{-}$ (Rodríguez and Fraga 1999; Alori et al. 2017). PSB increase solubilization of P in soil through many processes; for example, they may decrease the pH of the soil by the producing organic (gluconic acid) and mineral acids, alkaline phosphatases, phytohormones, and H^+ protonation, anion exchange, chelation, and siderophore production, which promotes P solubilization in soil (Mahdi et al. 2011; Ahemad 2015). The exploitation of these processes may prevent frequent addition of P into soil, with a substantial reduction in cost of production to farmers and damage to the environment. Among acids, gluconic acid is the most frequent organic acid produced by PSB by the enzyme glucose dehydrogenase's indirect oxidation pathway of glucose (Mahdi et al. 2011). Several reports have indicated that different bacterial species, particularly rhizobacteria, have the ability to liberate organic phosphates or to solubilize insoluble inorganic phosphate compounds such as dicalcium phosphate, tricalcium phosphate, hydroxyapatite, and rock phosphate (Chen et al. 2006a,b; Gulati et al. 2010; Al-Enazy et al. 2017; Xiao et al. 2017). These bacteria make available the soluble phosphates to plants and in return gain root-borne carbon compounds, mainly sugars and organic acids, necessary for bacterial growth (Rashid et al. 2004). PSB include free-living rhizobacteria, such as *Pseudomonas*, symbiotic nitrogen fixers (rhizobia), and asymbiotic nitrogen fixers (*Azotobacter*). The active strains of PSB involving insolubilization and mineralization include *Pseudomonas, Mycobacterium, Micrococcus, Bacillus, Flavobacterium, Rhizobium, Mesorhizobium*, and *Sinorhizobium* (Ahemad 2015). PSMs may enhance the phytoremediation competence of plants by promoting their growth and health, even in the presence of toxic metals.

6.2.2 Phosphate-Solubilizing Actinomycetes

The P-solubilizing ability of actinomycetes has attracted interest in recent years because this group of soil organisms is not only capable of surviving in extreme environments but also possesses other potential benefits (e.g., production of antibiotics and phytohormone-like compounds) that could simultaneously benefit plant growth (Alori et al. 2017). A study by Hamdali et al. (2008) has indicated that approximately 20% of actinomycetes can solubilize P, including those in the common genera *Streptomyces* and *Micromonospora*. A partial list of PSMs including various groups is given in Table 6.1.

6.2.3 Phosphate-Solubilizing Cyanobacteria

Cyanobacteria, also called blue green algae (BGA), are primary colonizers in different habitats. The practical utility of these organisms as a source of organic nitrogen fertilizer for rice has been well recognized; however, phosphorus is necessary for growth and nitrogen fixation by these organisms. Blue green algae, like PSB, are known to have the ability to mobilize bound phosphates. They have been shown to solubilize insoluble $Ca_3(PO_4)_2$, $FePO_4$, $AlPO_4$, and hydroxyapatite ($Ca_5(PO_4)_3 \cdot OH$) in soils, sediments, or pure cultures. They are also known to solubilize organic sources of phosphorus.

6.2.4 Phosphate-Solubilizing Fungi

The solubilization and mineralization of P by phosphate-solubilizing fungi are important traits of plant growth-promoting fungi such as mycorrhizae (Jones et al. 1991). Fungi in soils are able to traverse long distances more easily than bacteria and hence may be more important to P solubilization in soils. P-solubilizing fungi do not lose the P-dissolving activity upon repeated subculturing under laboratory conditions as occurs with P-solubilizing bacteria. Generally, P-solubilizing fungi produce more acids than bacteria and consequently exhibit greater P-solubilizing activity. *Aspergillus* and *Penicillium* are the most representative genera of P-solubilizing fungi. Strains of *Trichoderma* and *Rhizoctonia solani* have also been reported as potential P solubilizers. A list of PSMs including various groups is presented in Table 6.1.

6.3 MECHANISM OF PHOSPHATE SOLUBILIZATION

The solubilization of insoluble P by microorganisms was reported by Pikovskaya (1948). During the last two decades, knowledge of PSMs has increased significantly. Several strains of bacterial and fungal species have been described and investigated in detail for their phosphate-solubilizing capabilities (Glick 1995; Zhang et al. 2018). Typically, such microorganisms have been isolated using cultural procedures, with species of *Pseudomonas* and *Bacillus* bacteria (Illmer and Schinner 1995) and *Aspergillus* and *Penicillium* fungi being predominant (Wakelin et al. 2004). These organisms are ubiquitous but vary in density and mineral phosphate-solubilizing

TABLE 6.1
Examples of Potential Phosphate-Solubilizing Microorganisms

Category	Name of Microorganisms
Bacteria	*Alcaligenes* sp., *Aerobacter aerogenes*, *Achromobacter* sp., *Actinomadura oligospora*, *Agrobacterium* sp., *Azospirillum brasilense*, *Bacillus* sp., *B. circulans*, *B. cereus*, *B. fusiformis*, *B. pumilus*, *B. megaterium*, *Bacillus mycoides*, *B. polymyxa*, *B. coagulans*, *B. chitinolyticus*, *B. subtilis*, *Bradyrhizobium* sp., *Brevibacterium* sp., *Citrobacter* spp., *Pseudomonas* sp., *P. putida*, *P. fluorescens*, *P. calcis*, *Flavobacterium* sp., *Nitrosomonas* sp., *Erwinia* sp., *Micrococcus* sp., *Escherichia intermedia*, *Enterobacter asburiae*, *Serratia phosphoticum*, *Nitrobacter* sp., *Thiobacillus ferrooxidans*, *T. thiooxidans*, *Rhizobium meliloti*, *Xanthomonas* sp., *Agrobacterium* spp., *Azotobacter* spp., *Burkholderia* sp., *Enterobacter* sp., *Kushneria* sp., *Paenibacillus* spp., *Ralstonia* sp., *Rhizobium* spp., *Rhodococcus* sp., *Serratia* sp., *Salmonella* sp., *Sinomonas* sp., *Arthrobacter* sp., *Chryseobacterium* sp., *Gordonia* sp., *Phyllobacterium* sp., *Delftia* sp., *Azotobacter* sp., *Pantoea* sp., *Klebsiella*, *Vibrio proteolyticus*, *Xanthobacter agilis*, *Rhizobium leguminosarum* bv. Trifolii, and *Crotalaria* sp.
Fungi and Yeast	*Aspergillus awamori, A. niger, A. terreus, A. flavus, A. nidulans, A. foetidus, A. wentii, Fusarium oxysporum, Alternaria tenuis, Achrothcium* sp., *Penicillium digitatum, P. lilacinum, P. balaji, P. funiculosum, Cephalosporium* sp., *Cladosporium* sp., *Curvularia lunata, Cunninghamella, Candida* sp., *Chaetomium globosum, Humicola insolens, Humicola lanuginosa, Helminthosporium* sp., *Paecilomyces fusisporous, Pythium* sp., *Phoma* sp., *Populospora mytilina, Myrothecium roridum, Mortierella* sp., *Micromonospora* sp., *Oidiodendron* sp., *Rhizoctonia solani, Rhizopus* sp., *Mucor* sp., *Trichoderma viride, Torula thermophila, Schwanniomyces occidentalis, Sclerotium rolfsii, Achrothcium, Alternaria, Arthrobotrys* sp., *Aspergillus* sp., *Curvularia, Cunninghamella* sp., *Chaetomium, Fusarium, Glomus, Helminthosporium, Micromonospora, Mortierella, Myrothecium, Oidiodendron, Paecilomyces* sp., *Penicillium, Phoma, Pichia fermentans, Populospora, Rhizoctonia, Saccharomyces* sp., *Schizosaccharomyces, Schwanniomyces, Sclerotium* sp., *Torula, Trichoderma* sp., *Yarrowia* sp., *Arthrobotrys oligospora, Penicillium* sp., and *Penicillium aurantiogriseum, Yarrowia lipolytica, Schizosaccharomyces pombe*, and *Pichia fermentans*
Actinomycetes	*Actinomyces, Streptomyces albus, S. cyaneus*, and *Streptoverticillium album*
Cyanobacteria	*Anabaena* sp., *Calothrix braunii, Nostoc* sp., *Scytonema* sp., and *Westiellopsis prolifica*
Arbuscular mycorrhizae	*Glomus fasciculatum* and *Entrophospora colombiana*

ability from soil to soil or from one production system to another. The main P solubilization mechanisms employed by soil microorganisms include (i) release of complexing or mineral-dissolving compounds, for example, organic acid anions, siderophores, protons, hydroxyl ions, and CO_2; (ii) liberation of extracellular enzymes (biochemical P mineralization); and (iii) release of P during substrate degradation (biological P mineralization) (Rodríguez and Fraga 1999; Mahdi et al. 2011; Ahemad

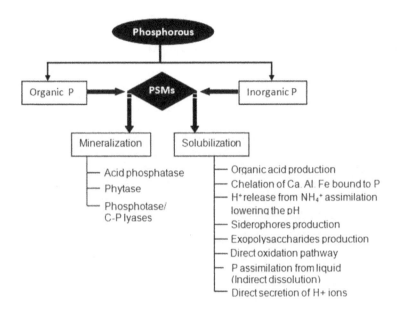

FIGURE 6.1 Schematic representation of the mechanism of soil P solubilization/ mineralization by phosphate-solubilizing microorganisms (PSMs).

2015). On the basis of mode of action by microorganisms, the mechanism of P solubilization can be divided into two categories (i) inorganic P solubilization and (ii) organic P solubilization (Figure 6.1).

6.3.1 Inorganic P Solubilization

The monovalent anion phosphate $H_2PO_4^-$ is a major soluble form of inorganic phosphate, which usually occurs at lower pH values. However, as the pH of the soil environment increases, the divalent and trivalent forms of Pi (HPO_4^{-2} and HPO_4^{-3}, respectively) occur. Typically, the solubilization of inorganic P occurs as a consequence of the action of low-molecular-weight (LMW) organic acids such as gluconic and citric acid, both of which are synthesized by various soil bacteria (Zheng et al. 2018). Thus, the extrusion of LMW organic acids by the PSM strains into the surrounding environment acidifies the cells and their surrounding environment and ultimately leads to the release of P ions from the P mineral by anion exchange of phosphate by acid anions or by forming soluble complexes with some cations, including Ca^{2+}, Fe^{3+}, and Al^{3+} associated with insoluble P via hydroxyl and carboxyl groups, and thus P is released in soils. To explain the previous mechanism of P solubilization, several theories have been proposed by various groups of researchers, including (i) the sink theory (Halvorson et al. 1990), (ii) the organic acid theory, and (iii) the acidification by H^+ excretion theory (Illmer and Schinner 1995). However, the organic acid theory is well recognized and the most widely accepted mechanism of P solubilization by PSMs. Accordingly, the prominent acids released by PSMs in the solubilization of insoluble P are gluconic acid, oxalic acid, citric acid, lactic acid, tartaric acid, and aspartic acid. These organic acids drop the pH to bring P into solution or chelate mineral ions. This, in turn, leads to acidification of microbial cells and their surroundings and, consequently, the release of P ions from the P mineral by H^+ substitution for Ca^{2+}. However, acidification does not seem to be the only mechanism of solubilization, as the ability to reduce the pH in some cases did not correlate with the ability to solubilize mineral P). Altomare et al. (1999) investigated the capability of the plant growth-promoting and biocontrol fungus *Trichoderma harzianum* T-22 to solubilize *in vitro* insoluble minerals, including rock phosphate. Organic acids were not detected in the culture filtrates and hence the authors concluded that acidification was probably not the major mechanism of solubilization, as the pH never fell below 5.0. The solubilization of inorganic P, however, is complex and depends on numerous factors such as soil properties, plant nutritional requirements, and physiological and growth conditions.

6.3.2 ORGANIC P SOLUBILIZATION

In soil, the major source of organic P is organic matter, and thus the amount of organic P can be as high as 30%–50% of the total P. Soil organic P largely exists in the form of inositol phosphate (soil phytate). Other organic P compounds that have been reported are phosphomonoesters, phosphodiesters, phospholipids, nucleic acids, and phosphotriesters. Organic P solubilization is called mineralization of organic P, in which P plays an imperative role in P cycling of a farming system. Organic P can be released from organic compounds in soil by the action of several enzymes such as phytases, C-P lyases, and phosphatases. The phytase enzyme is responsible for the release of P from organic materials in soil (plant seeds and pollen) that are stored in the form of phytate. Degradation of phytate by phytase releases P in a form that is available for plant use. Hence, microorganisms are in fact a key driver in regulating the mineralization of phytate in soil, and their presence within the rhizosphere may compensate for a plant's inability to otherwise acquire P directly from phytate. Plants generally cannot acquire P directly from phytate; however, the presence of PSMs within the rhizosphere may compensate for a plant's inability to otherwise acquire P directly from phytate. Mineralization of most organic P compounds is carried out by means of phosphatase enzymes. The major source of these enzymes in soil is considered to be of microbial origin.

6.4 MECHANISMS OF PHOSPHATE-SOLUBILIZING MICROBES IN PHYTOREMEDIATION OF HEAVY METALS

Natural solubilization of mineral phosphates is an important phenomenon exhibited by different PSMs. In addition to P solubilization, PSMs not only protect plants from phytopathogens through the production of antifungal metabolites (such as phenolics and flavonoids), antibiotics, hydrogen cyanide (HCN), phenazines, and

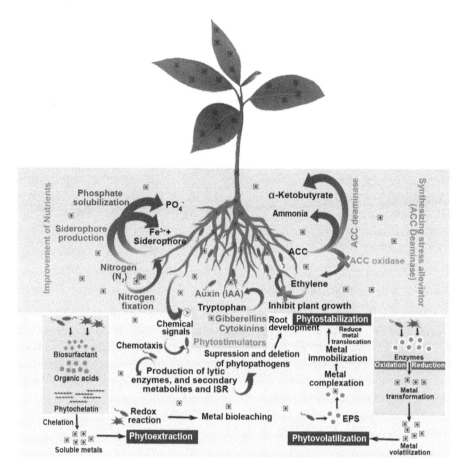

FIGURE 6.2 Effects of the functioning of phosphate-solubilizing microorganisms on the biogeochemical cycling of heavy metals and their implication for phytoremediation.

lytic enzymes but also promote plant growth through biological nitrogen fixation (BNF), production of metal-accumulating siderophores, and phytohormone secretion (Rodríguez Fraga 1999; Rashid et al. 2004; Mahdi et al. 2011; Glick 2012; Kalayu 2019). In addition, various plant growth-promoting traits of PSMs, such as secretion of siderophores, organic acid production, IAA production, and ACC deaminase activity, contribute to enhancing the phytoremediation capability of plants in a polluted environment. Various plant growth-promoting (PGP) traits of PSMs facilitate remediation of HMs through a variety of direct or indirect mechanisms (Kuffner et al. 2010; Kong and Glick 2017). A heavy metal–resistant bacterium can alleviate the stress of heavy metals in plants using one or more of the mechanisms mentioned previously. Figure 6.2 illustrates mechanisms used by plant growth-promoting traits of PSMs to improve the phytoremediation potential of plants.

6.4.1   Direct Mechanisms

The direct mechanisms involved in plant-growth promotion in HM-contaminated environments include phytostimulation, 1-aminocyclopropane-1-carboxylate deaminase activity, BNF, nutrient solubilization/mobilization, and siderophore production. PGPR is also effective in improving plant growth in stress conditions through ACC-deaminase activity, exopolysaccharide (EPS) production, and scavenging toxic reactive oxygen species (Glick 2005, 2010, 2012).

6.4.1.1   Phytostimulation

Phytostimulation is the direct promotion of plant growth and development processes, for example, apical dominance, cell growth, tropisms, initiation of adventitious and lateral roots, cell and vascular differentiation, stamen development, and biotic and abiotic stress resistance, through the production of phytohormones. Several PSMs synthesize significant quantities of phytohormones such as indole-acetic acid (IAA), gibberellins, auxins, brassinosteroids (BRs), cytokinins, abscisic acid, and ethylene and release them into the plant, resulting in pronounced positive effects on plant growth and development. These phytohormones play a significant role in adapting plants to HM toxicity stress through mediating a wide range of adaptive responses (Peleg and Blumwald 2011). Among these phytohormones, the most highly studied example of phytostimulation involves lowering plant hormone ethylene levels by the enzyme ACC deaminase that converts the ethylene precursor ACC to α-ketobutyrate and ammonia and promotes plant growth, especially during stress conditions, by reducing the level of stress ethylene to below the point where it is inhibitory to growth. Ethylene, also known as a stress hormone, is an important phytohormone in modulating the growth and cellular metabolism of plants, but the overproduction of ethylene promoted by stresses can inhibit plant developmental processes, such as root elongation, lateral root growth, and root hair formation. Under stress conditions, including drought, ethylene endogenously regulates plant homeostasis and results in reduced root and shoot growth. However, degradation of the ethylene precursor ACC by bacterial ACC deaminase releases plant stress and rescues normal plant growth. Moreover, a number of studies have shown that ACC deaminase containing PSMs can facilitate metal phytoremediation by reducing the level of stress ethylene that is induced by high levels of toxic metals.

6.4.1.2   Nutrient Solubilization/Mobilization

Soil microorganisms are solely responsible for nutrient cyclings such as N, P, S, and other nutrients. HMs in the rhizosphere affect nutrient uptake, which retards plant growth. Under such nutrient-limiting conditions, PGP bacteria have the potential to provide essential nutrients to plants. In addition to the decomposition of soil organic matter, microbes also make chemically fixed nutrients, such as phosphorus (P), zinc (Zn), potassium (K), and Fe, available. The main mechanism in the solubilization of P, K, Fe, and Zn is the lowering of pH from the production of organic acids. Wu et al. (2006b) studied soil bacteria that were resistant to Zn, Cu, and Pb and found that a P-solubilizer (*Bacillus megaterium* HKP-1) and a K-solubilizer (*B. mucilaginosus* HKK-1) had a high resistance to HM toxicity (Zn, Cu, and Pb) because of the formation of endospores, which are capable of withstanding extreme conditions. Mumtaz et al. (2017) also demonstrated Zn solubilization by screening 70 isolates and identified the four best ZSB strains, *Bacillus* sp. ZM20, *B. aryabhattai* ZM31, *B. subtilis* ZM63, and *B. aryabhattai* S10. In soil, siderophore production activity plays a central role in determining the ability of different microorganisms to improve plant development. Microbial siderophores enhance iron uptake by plants that are able to recognize the bacterial ferric-siderophore complex and are also important in iron uptake by plants in the presence of other metals such as nickel and cadmium (Burd et al. 1998; Dimkpa et al. 2008). However, it is still unclear if bacterial siderophore complexes can significantly contribute to the iron requirements of the plant. Examples of successful remediation of heavy metals by using plant-rhizosphere bacteria partnerships are listed in Table 6.2.

6.4.1.2.1   Organic Acids

In heavy metal–contaminated soils, metals strongly adhere to soil particles; therefore, they are not easily available for uptake by phytoextracting plants. In this context, PSB are very promising agents since they solubilize insoluble and biologically unavailable metals such as Ni (Becerra-Castro et al. 2009), Cu (Li and Ramakrishna 2011), and Zn (He et al. 2013) by secreting LMW organic acids; thus, they facilitate metal bioavailability for plant uptake (Becerra-Castro et al. 2009). Organic acids, including acetate; lactic, citric, 2-ketogluconic, malic, glycolic, oxalic, malonic, tartaric, valeric, succinic, and formic acids; lactate; malate; oxalate; succinate; citrate; gluconate; ketogluconate; and so on, can form complexes with the iron or aluminum in ferric and aluminum phosphates, thus releasing plant-available phosphate into the soil. For example, canola plants inoculated with the plant growth-promoting endophytic bacterium *Rahnella* sp. JN6 had significantly greater biomass, concentrations, and uptake of Cd, Pb, and Zn in both aerial and root tissues than those without inoculation grown in heavy metal–amended soils (He et al. 2013). Nevertheless, the observed increases in both plant biomass and metal uptake were attributed to phosphate solubilization as well as other plant growth-promoting traits, such as ACC deaminase activity, IAA production, and siderophore synthesis. It has been proven that organic acid-producing PGPR can alleviate metal-induced stress in plants and help survival and growth of plants grown in heavy metal–polluted soils by forming complexes with heavy metals that are less phytotoxic than the free forms of heavy metals for plants.

6.4.1.2.2   Siderophores

The supply of Fe to growing plants under heavy metal pollution becomes more important as bacterial siderophores help to minimize the stress imposed by metal contaminants. Studies have reported the release of siderophores from PSMs (Vassilev et al. 2006; Hamdali et al. 2008a); however, siderophore production has not been widely implicated as a P solubilization mechanism. Microbially released siderophores increase plant Fe uptake through different mechanisms, such as chelation and release of Fe, direct uptake of Fe-complexed siderophores, or ligand exchange reactions. As a result, the soluble metal concentration increases through binding with siderophores. Inoculation with siderophore-producing rhizobacteria improved plant growth and nutrient assimilation (Rajkumar et al. 2010). Under low Fe conditions, siderophores

TABLE 6.2

Summary of Reports on the Role of Phosphate-Solubilizing Bacteria in Plant-Growth Promotion and Remediation of Heavy Metal–Contaminated Soil

Bacteria	Host Plant	Pollutants Uptake	Plant-Growth Promotion Features	Reference
Achromobacter xylosoxidans Ax10	*Brassica juncea*	Cu	ACCD, IAA, P solubilization	Ma et al. (2009b)
Arthrobacter sp. MT16, *Microbacterium* sp. JYC17, *Pseudomonas chlororaphis* SZY6, *Azotobacter vinelandii* GZC24, *Microbacterium lactium* Y17	*Brassica napus*	Cu	ACCD, siderophores, IAA, P solubilization	He et al. (2010a)
Bacillus mucilaginosus HKK-1, *Bacillus megaterium* HKP-1	*Brassica juncea*	Zn, Pb, and Cu	P, K solubilization	Wu et al. (2006b)
Bacillus thuringiensis GDB-1	*Alnus firma*	As	ACCD, IAA, siderophores, P solubilization	Babu et al. (2013)
Burkholderia sp. J62	*Zea mays, Lycopersicon esculentum*	Pb and Cd	IAA, siderophores, ACCD, P solubilization	Jiang et al. (2008)
Enterobacter intermedius MH8b	*Sinapis alba*	Zn	ACCD, IAA, hydrocyanic acid, P solubilization	Plociniczak et al. (2013)
Enterobacter sp. JYX7, *Klebsiella* sp. JYX10	*Polygonum pubescens*	–	IAA, siderophores, ACCD, P solubilization	Jing et al. (2014)
Phyllobacterium myrsinacearum RC6b	*Sedum plumbizincicola*	Cd, Zn, and Pb	ACCD, IAA, siderophores, P solubilization	Ma et al. (2013)
Pseudomonas aeruginosa MKRh3	*Vigna mungo*	–	ACCD, siderophores, IAAsynthesis, P solubilization	Ganesan (2008)
Pseudomonas sp. M6, *Pseudomonas jessenii* M15	*Ricinus communis*	Ni, Zn, and Cu	IAA, ACCD, P solubilization	Rajkumar and Freitas (2008)

(Continued)

TABLE 6.2 (Continued)

Summary of Reports on the Role of Phosphate-Solubilizing Bacteria in Plant-Growth Promotion and Remediation of Heavy Metal–Contaminated Soil

Bacteria	Host Plant	Pollutants Uptake	Plant-Growth Promotion Features	Reference
Rahnella sp. JN27	Amaranthus hypochondriacus, A. mangostanus, Solanum nigrum	Cd	IAA, siderophores, ACCD, P solubilization	Yuan et al. (2014)
Bacillus aryabhattai		–	Zn solubilization	Ramesh et al. (2014)
Bacillus sp. ZM20, Bacillus aryabhattai ZM31, Bacillus subtilis ZM63, Bacillus aryabhattai S10		–	Zn solubilization	Mumtaz et al. (2017)
Burkholderia metalliresistens sp. nov., D414T	–	Pb, Cd, Cu, and Zn	P solubilization	Guo et al. (2015)

Abbreviations: P, Phosphate; ACCD, 1-aminocyclopropane-1-carboxylic acid (ACC) deaminase; IAA, Indole-acetic acid.

can solubilize Fe from minerals and organic compounds (Khan et al. 2009a). Siderophores also form stable complexes with heavy metals, alleviating the toxic effect of heavy metals (Rajkumar et al. 2010). Marathe et al. (2017) inoculated *Glycine max* seeds with siderophore-producing *Pseudomonas aeruginosa* 6a(bc4) and reported enhanced growth. Inoculation also showed antifungal activity against *Aspergillus* spp. In another study, Sharma et al. (2003) inoculated mung bean (*Vigna radiata* L. Wilzeck) with a *Pseudomonas* sp. strain GRP$_3$ strain with the ability to produce siderophores and reported enhanced Fe contents in inoculated plants. Rajkumar et al. (2010) reported the uptake of Fe in cereal grains through production of siderophores by rhizobacteria. *P. fluorescens* synthesizes Fe pyoverdine, which can increase Fe uptake in *Arabidopsis thaliana* tissues, enhancing plant growth (Vansuyt et al. 2007). In general, some of the known mechanisms by which siderophore-producing PGPRs could alleviate metal-induced stress in plants and help survival and growth of plants in heavy metal–contaminated soils include providing nutrients, particularly Fe (improvement of Fe availability), to metal-stressed plants; reducing uptake of HMs (e.g., Cd, Ni, and Pb); free radical formation around plant roots and thus reduction oxidative stress (Dimkpa et al. 2009).

6.4.1.3 Biological Nitrogen Fixation

Nitrogen (N) is one of the main components of proteins, nucleic acids, vitamins, and hormones. BNF, the ability to reduce atmospheric N$_2$ gas to biologically available ammonium (NH$_4^+$), is a process performed by a diverse array of prokaryotic microorganisms collectively known as diazotrophs. Examples are NFB (such as rhizobia, rhizobacteria, and endophytic bacteria) and nutrient-absorbing arbuscular mycorrhizal fungi (AMF), which can improve the fertility of polluted soils for plant growth by catalyzing the reduction of atmospheric N$_2$ to NH$_4^+$. Microbe-mediated mineralization of N to NH$_4^+$ and its subsequent nitrification to NO$_3^-$ is of major significance to N availability (Figure 6.1). Diazotrophic phosphate-solubilizing bacteria include *Rhizobium, Bacillus,* and *Pseudomonas* and can fix atmospheric nitrogen, even under metal-stressed conditions.

6.4.1.4 Production of Biosurfactants

Biosurfactants, surface-active substances, are amphiphilic compounds that are either produced on the surfaces of the microbial cell or excreted extracellularly and reduce surface and interfacial tensions. They are usually composed of one or more compounds such as glycolipids, fatty acids, phospholipids, lipoproteins or lipopeptides, and mycolic acid (Pacheco et al. 2010). Bacterial biosurfactants increase HM tolerance and aid in metal removal from soil (Pacwa-Płociniczak et al. 2011). Due to their amphiphilic structure, these substances are able to create complexes with HMs at the soil interface and desorb HMs from the soil matrix to the soil solution, thus increasing metal bioavailability and solubility in metal-contaminated soils (Gamalero and Glick 2012; Singh and Cameotra 2013). As an example, the use of bacterially produced di-rhamnolipid surfactant has been shown to alleviate metal stress in Cd- and Pb-contaminated soils by removing 92% Cd and 88% Pb (Juwarkar et al. 2007). Some researchers also reported that HMs were removed from contaminated soils by biosurfactants of sophorolipids, di-rhamnolipids, and rhamnolipids produced by

B. subtilis, Torulopsis bombicola, and *P. aeruginosa* (Juwarkar et al. 2007; Venkatesh and Vedaraman 2012).

6.4.2 INDIRECT MECHANISMS

The indirect promotion of plant growth occurs when PSMs lessen or prevent the deleterious effects of one or more phytopathogenic organisms, usually fungi and nematodes. This can happen by producing antagonistic substances or by inducing resistance to pathogens (Glick 1995). A particular PSM, especially PGPR, may affect plant growth and development by using any one, or more, of these mechanisms. The ability of plant growth-promoting bacteria to act as biocontrol agents against phytopathogens and thus indirectly stimulate plant growth may result from any one of a variety of mechanisms, including antibiotic production, depletion of iron from the rhizosphere, induced systemic resistance, production of fungal cell wall–lysing enzymes, and competition for binding sites on the root (Glick 1995).

6.4.2.1 Microbial Antagonism

According to Beattie (2006), microbes that exhibit antagonistic activity toward a pathogen are defined as antagonists. The following rhizospheric environment and bacterial antagonistic activities can be highlighted: (i) synthesis of hydrolytic enzymes, such as chitinases, glucanases, proteases, and lipases, that can lyse pathogenic fungal cells; (ii) competition for nutrients and suitable colonization of niches at the root surface; (iii) regulation of plant ethylene levels through the ACC-deaminase enzyme; and (iv) production of siderophores and antibiotics. Antibiotics identified in antagonistic PGP biocontrol bacteria include the classical compounds HCN; phenazines, of which the major ones are carboxamide; 2,4-diacetylphloroglucinol (Phl); pyoluteorin; pyrrolnitrin, tensin, and viscosinamide. Zwittermicin A and kanosamine can be produced by *Bacillus cereus*. Some metal-resistant PSMs have also been reported to be able to produce enzymes such as chitinase, β-1,3 glucanase, protease, or lipase, by which they can lyse the cells of fungal pathogens (van Loon et al. 2007).

6.4.2.2 Induced Systemic Resistance

Induced systemic resistance (ISR) is a plant defense mechanism that is activated against infection by phytopathogens, including fungal, bacterial, and viral pathogens, as well as nematodes and insects. In ISR, interaction of some PSMs with plant roots can result in plants resistant to phytopathogenic attack. ISR shares many properties with innate immunity in humans. ISR was formerly described by Van Peer et al. (1991) in carnation plants that were systemically protected by the *P. fluorescens* strain WCS417r against *Fusarium oxysporum* f. sp. dianthi and by Wei et al. (1991) in cucumber plants, where rhizobacterial strains protected the leaves against anthracnose caused by *Colletotrichum orbiculare*. Specifically, *Pseudomonas* and *Bacillus* spp. are the rhizobacteria most studied that trigger ISR (Kloepper et al. 2004). Rhizobacteria-mediated ISR resembles pathogen-induced systemic acquired resistance (SAR) in that both types of induced resistance render uninfected plant parts more resistant to plant pathogens. Vleesschauwer and Höfte (2009) proposed the terminology ISR to depict induced systemic resistance promoted by non-pathogenic rhizobacteria or PGPR, irrespective of the signaling pathway involved in this process, while the term SAR is used to describe salicylic acid–dependent induced resistance triggered by a localized infection

6.4.2.3 Competition for Ferric Iron Ions

The siderophores produced by PSMs, such as bacteria and fungi, under iron-limiting conditions protect the plant from phytopathogens by chelating iron ions (Fe^{3+}) in the rhizosphere, which can be taken up by plant roots as an Fe nutrient and thus reduce its availability to pathogens that are reliant on available iron in soil (Rajkumar et al. 2010). Siderophores are produced by a diverse group of microorganisms; however, their production is more common among PGPB that exhibit optimum growth and siderophore production activity at extreme environmental conditions, including the scarcity of nutrients or the presence of elevated concentrations of heavy metals (Rajkumar et al. 2010). Siderophores produced by biocontrol PGPB have a higher affinity for iron than other pathogens like fungal pathogens; due to this, fungal pathogens cannot proliferate in the root zone of plants because of low levels of iron. Some studies show that fungal pathogens may suppress the siderophores of biocontrol PGPB and cause many plant diseases. For example, in some studies, mutants were used that were defective and unable to produce siderophores and found to be less effective against fungal pathogens than the wild type at protecting plants (Martinetti and Loper 1992; Buysens et al. 1996).

6.4.2.4 Competition for Nutrients and Colonization of Niches

Although it is difficult to demonstrate directly, some indirect evidence indicates that competition between pathogens and nonpathogens (PGPB) can limit disease incidence and severity. Thus, for example, abundant nonpathogenic soil microbes rapidly colonize plant surfaces and use most of the available nutrients, making it difficult for pathogens to grow. For example, in one series of experiments, researchers demonstrated that the treatment of plants with the leaf bacterium *Sphingomonas* sp. prevented the bacterial pathogen *Pseudomonas syringae* pv. tomato from causing pathogenic symptoms.

6.4.2.5 Predation and Parasitism

Predation and parasitism, the major biocontrol mechanisms used by some (fungal) *Trichoderma* species, are based on enzymatic destruction of the fungal cell wall. To our knowledge, this mechanism has not thoroughly been identified in bacteria.

6.4.2.6 Stress Controllers

Plants typically respond to the presence of phytopathogens by synthesizing stress ethylene that exacerbates the effects of the stress on the plant. Therefore, one way to decrease the damage to plants caused by a wide range of phytopathogens is to lower the plant's ethylene response. The simplest way to do this is to treat plants (generally the roots or seeds) with ACC deaminase containing PGPB. ACC deaminase–containing bacteria can lower plant ethylene levels in plants and thereby provide some protection against the inhibitory effects of various stresses (Glick 2005).

6.5   MODE OF ACTION OF PHOSPHATE-SOLUBILIZING MICROBES IN REDUCING HEAVY METAL UPTAKE

It has been reported that some HM-resistant PSMs can also reduce plant metal uptake or translocation to aerial plant parts by decreasing metal bioavailability in soil by processes of bioaccumulation, biosorption, precipitation, biotransformation (via methylation, demethylation, volatilization, complex formation, oxidation, or reduction), complexation, and alkalization (Ahemad 2015; Ma et al. 2016a). Bioaccumulation and biosorption are among the mechanisms used by HM-resistant PSMs to prevent the absorption of heavy metals by plants grown in HM-contaminated soils. In bioaccumulation, microbes retain and concentrate HMs in their bodies. Bioaccumulating microbial strains can be strong candidates for decontamination of polluted soil and water, as reported by Akhter et al. (2017). They isolated and identified three bacterial strains of *Bacillus cereus,* that is, BDBC01, AVP12, and NC7401, from the rhizosphere of *Tagetes minuta* and reported that these strains have strong solubilization and accumulation potential for Cr^{6+}, Ni^{2+}, and Cd^{2+} and thus help in biosorption of these metals. Biosorption of HMs is the sequestration of positively charged metal ions by ionic groups on cell surfaces (Malik 2004).The bioaccumulation process is more complex than biosorption and requires metabolic activity of living cells involving intracellular sequestration (MT and PC binding), extracellular precipitation, metal accumulation, and complex formation (Gadd 2004). Certain plant-associated microorganisms have the ability to promote the enzymatically catalyzed precipitation of radionuclides (e.g., U, Tc) and toxic metals (e.g., Cr, Se) by microbial reduction processes, which show considerable promise for phytoremediation of metal-contaminated soils (Payne and DiChristina 2006). Oves et al. (2013) reported that the inoculation of Cr-reducing bacterium *P. aeruginosa* OSG41 onto chickpea grown in Cr^{6+}-contaminated soils significantly decreased Cr uptake in roots, shoots, and grains, with a concomitant increase in plant growth performance compared with noninoculated control. Also, insoluble mineral forms of metals can also be immobilized, either directly through an enzymatic action or indirectly by bacterial Fe oxidation or interactions between microbial inorganic acids (e.g., hydrogen sulfide, bicarbonate, and phosphate). By chelating metal ions, natural organic chelating agents such as biosurfactants, metallophores, organic acid anions, and siderophores can decrease metal bioavailability and toxicity in the rhizosphere (Payne and DiChristina 2006; Majumder et al. 2013; Sessitsch et al. 2013). Microorganisms can mobilize metals (Pb, Hg, Se, As, Tn, and Sn) through biomethylation, which can result in volatilization (Gadd 2004; Barkay and Wagner-Dobler 2005; Bolan et al. 2014). Volatile methylated species are often lost from the soil. Organic matter in complexation with iron minerals helps the adsorption of metals in soil environments. The excretion of EPS by plant-associated microbes is of particular importance to form a protective barrier against harmful effects through metal biosorption (Slaveykovaetal 2010; Hou et al. 2015). The mechanisms involved in metal biosorption onto EPS include metal ion exchange, complexation with negatively charged functional groups, adsorption, and precipitation.

7 Phytoaugmentation
An Emerging and Sustainable Approach for Remediation of Contaminants in Wastewater through Augmenting Phytoremediation Technology

7.1 INTRODUCTION

Indiscriminate and uncontrolled discharge of industrial, as well as domestic and municipal, wastes into the environmental sink has become an issue of major global concern (Chandra et al. 2018a,b). Different industries such as pulp and paper mills, distilleries, tanneries, textiles, pharmaceuticals, food, and petroleum processing industries discharge very complex wastewater into water bodies, adversely affecting humans, animals, and normal operations of ecosystems and creating health hazards (Khandare et al. 2013; Pei et al. 2016; Ashraf et al. 2018a,b; Afzal et al. 2019; Kumar et al. 2020). Various physico-chemical methods such as flocculation, coagulation, adsorption, membrane filtration, chemical precipitation, ion-exchange, UV/H_2O_2 treatment, and ozone oxidation have been reported for degradation and decolorization of complex effluent, but these techniques are not feasible at large scale due to high cost, generation of huge amounts of toxic sludge, and other secondary pollutants (Kabra et al. 2013; Hussain et al. 2018). On the other hand, there are numerous conventional biological treatment processes such as activated sludge and anaerobic digestion processes that are cheaper than physico-chemical methods that are widely used in the treatment of complex industrial wastewater. However, there are still some problems in applying these approaches to real wastewater treatment due to the death of microbial cells in high concentrations of toxic organic and inorganic pollutants. Biological processes are usually aimed at removing readily biodegradable organic compounds or relatively readily hydrolysable substances but are not specifically designed to remove heavy metals from complex wastewater (Chandra and Kumar 2017a,b,c). In addition, continuous supplies of expensive nutrients or chemicals create difficulty in the separation of cells after treatment. In such cases, the wastewater treatment process requires several treatment steps in order to produce a treated effluent with the required quality to be discharged in the natural receiving waters. On the other hand, phytoremediation has offered a promising approach for sustainable management of industrial wastewater or wastewater-contaminated soil and water (Salt et al. 1995; Dickinson et al. 2009). Phytoremediation is one of the bioremediation technologies that use plants and their root-associated microbes for environmental cleanup and is generally considered an economical, environmentally friendly and aesthetically acceptable approach to treating toxic wastewater (Ashraf et al. 2019). However, phytoremediation of numerous organic contaminants often proves not ideal because plants and their associated microbes only partially degrade these compounds. In addition, some of the pollutants are toxic enough to downgrade the overall cleaning process by reducing plant growth and inhibiting microbial degradation (Rajkumar et al. 2012; Yadav et al. 2014). To overcome this situation, phytoaugmentation, the introduction of natural or genetically modified pollutants degrading rhizospheric and/ or endophytic microorganisms with plants to enable or accelerate degradation of contaminants, is a promising approach for the treatment of hazardous wastewater pollutants and to meet effluent standards (Redfern and Gunsch 2016).

This chapter presents insights into the role of plant–bacteria synergism in the removal/degradation of pollutants from wastewater in constructed and floating treatment wetlands and the factors affecting treatment performance.

7.2 PHYTOREMEDIATION VERSUS PHYTOAUGMENTATION

Phytoremediation has emerged as an energy-efficient and eco-friendly remediation technology for contaminated environments. It is a set of techniques emphasizing the efficient use of plants and their related enzymes and associated microbes for sequestration, detoxification, and mineralization of toxicants through complex natural biological, physiological, and chemical processes and activities of plants and microbes both (Kumar and Chandra 2018a). Restoration of habitats and *in situ* cleanup of contaminants can be achieved with significantly reduced remedial costs by this plant-microbe–based phytotechnology (Phieler et al. 2014). Plants can uptake contaminants persisting in the environment through root systems, which, by providing a larger surface area, facilitates mobilization, cleanup, or detoxification of contaminants within plants through various mechanisms, that is, elimination, containment, degradation, and so on. These plant properties have been used for effective elimination of waste, including metals, phenolic compounds, azo dyes and colorants, and various refractory organic and inorganic contaminants (Chandra et al. 2018a,b). Hence, depending upon the detoxification process, applicability, medium, and type and extent of pollution, phytoremediation processes can be subdivided into: *phytoextraction:* employs metal-accumulating plants that take up toxic metals from contaminated media and concentrate them in the harvestable biomass; *rhizofiltration:* absorption or adsorption of the contaminant by the plant root; *phytodegradation:* absorption and uptake of pollutants by the root system, resulting in metabolic or enzymatic transformation within or external to plants; *rhizodegradation:* secretion of root exudates or enzymes in the rhizosphere and subsequent microbial degradation of contaminants; and *phytovolatilization:* involves the plant uptake of metals from soil and their volatilization from the foliage. These plants can subsequently be harvested, processed, or disposed of safely. Although phytoremediation is an efficient and eco-friendly remediation technology for the decontamination of surface and groundwater or other polluted media, the success of these processes often depends on microbial action to modify and/ or degrade the contaminant (Kumar and Chandra 2018a). Since the appropriate microorganisms may not be present at a given site, phytoremediation of numerous organic contaminants often proves not ideal because plants and their root-associated microorganisms only partially degrade these toxic compounds (Ma et al. 2016a,b). To overcome this situation, researchers have investigated the inoculation of natural or genetically modified pollutant-degrading microbial strains or mixes of strains with plants to enable or accelerate microbial activities as well as plant metabolism in removing targeted pollutants and to meet effluent standards (Runes et al. 2001; Tara et al. 2019). Phytoaugmentation is considered a promising technology for the remediation of industrial and domestic effluent as well. It is considered a relatively cost-effective and environmental friendly technology compared to physico-chemical and conventional biological treatment methods to remediate environments polluted with pesticides and aromatic compounds and wastewater contaminated with phenol, lignin, melanoidins, phenanthrene, nitrobenzene, phenol dye, and so on (Huertas et al. 2006; Shao et al. 2014; Redfern and Gunsch 2016; Tara et al. 2019).

7.3 BIOLOGY AND FUNCTION OF PHYTOAUGMENTATION TECHNOLOGY: PLANT–BACTERIA SYNERGISM

Phytoaugmentation technology using constructed wetlands (CWs) and floating treatment wetlands (FTWs) gives freedom from employing expensive field instrumentation, rendering it sustainable and environmentally friendly; hence, it allows the treatment of domestic and/or industrial effluent at successful rates, particularly for countries with more economic constraints (Kadlec et al. 2000; Vymazal et al. 2006, 2011; Afzal et al. 2019; Tara et al. 2019). This technology exploits plants, bacteria, and their synergism for the remediation of contaminated soil and water (Kadlec and Knight 1996). Plants, especially macrophytes, being active partners in such a synergistic interaction, perform effective phytoremediation on their own, that is, phytoextraction, phytovolatilization, rhizofiltration, and rhizoremediation, and provide metabolites, nutrients, and habitat to microbes in their rhizosphere and endosphere (Brix and Schierup 1989; Chandra and Kumar 2015b). In plant–bacteria consortia, plants provide nutrients to rhizospheric microbes through their root exudates, and, in return, microbes improve plant growth and protect the plant through pollutant detoxification (Glick 2003, 2012). Microorganisms (rhizospheric and endospheric) play an important role in the mineralization of organic pollutants and the biogeochemical transformation of nutrients (i.e., assimilating nitrogen [N] and phosphorous [P] and concentration of nitrate [NO_3^-], phosphate [PO_4^{2-}], and heavy metals in wastewater through metabolism-dependent and independent methods in wetlands [Reddy et al. 2002]). The immersed rhizomes and roots of macrophytes provide a large surface area to develop a microbial biofilm, which plays a key role in the entrapment of fine suspended particles and consumption of excess nutrients in the water column (Headley and Tanner 2012). These microorganisms have major roles in the degradation and transformation of organic compounds, and their efficiency in degrading organic pollutants increases in the presence of plants (Glick 2003; Stottmeister et al. 2003; Chandra and Kumar 2015b). Moreover, the roots of some plants supply dissolved oxygen into the water body, benefiting the growth of aerobic microorganisms, which break down organic substances (Reddy et al. 2002). Some root-colonizing bacteria penetrate the root, colonize within it, and/or migrate to the aerial parts; these are known as endophytes (Doty 2008). Among endophytes, endophytic bacteria, however, have received more attention in the context of phytoremediation, as they develop an intimate relationship with the host as compared to the rhizospheric bacterial communities (Hussain et al. 2018). It has been shown that endophytic bacteria possess pollutant-degrading genes containing genes encoding for enzymes responsible for and to reduce both evapotranspiration and toxicity of volatile organic and inorganic pollutants (Weyens et al. 2010a). Furthermore, the use of endophytes with plant growth–stimulating activities may improve the plant's survival and development in a polluted environment. The microbial population on the root surface and inside the plant tissues enhances the removal of pollutants from water and wastewater. Once the pollutant is taken up by the plant, endophytes are actively involved in the *in planta* degradation and detoxification (Weyens et al. 2009a,c). As endophytes reside in the plant, they can interact more closely with their host plant as compared to rhizobacteria. There is increasing evidence that plant–microbe interactions/dynamics can define the efficiency of metal phytoextraction. Inoculation of the plant rhizosphere with microorganisms is an established route to improve phytoextraction efficiency by increasing the bioavailability of metal(loid)s to the plant and increasing plant biomass (Glick 2012). Selection of the appropriate plant to stimulate contaminant degradation is a key aspect in phytoremediation technology; however, for phytoaugmentation strategies, the choice of the contaminant-degrading microbes used is also important. Numerous factors, including pH of water and sediment, mobilization and uptake from the soil, compartmentalization and sequestration within the root, efficiency of xylem loading and transport (transfer factors), distribution between metal sinks in the aerial parts, sequestration and storage in leaf cells, and plant growing and transpiration rates, can also affect remediation processes of contaminated sites. These processes are generally directly and/or indirectly influenced by the different loading rates, temperatures, soil types, operation strategies, and redox conditions in the wetland bed.

7.3.1 Constructed Wetland: A Complex Bioreactor for Wastewater Treatment

CWs, also known as treatment wetlands, are effective, dynamic, and variable engineered systems designed and constructed to utilize natural processes involving wetland vegetation, soils, and their associated microbial assemblages to eradicate a wide variety of contaminants, namely suspended solids, organic compounds, nutrients, pathogens, metals, and emerging contaminants from wastewater within a more controlled environment (Kadlec and Wallace 2008; Vymazal 2011). CWs have received great attention from both scientists and engineers in the last decades and offer an effective alternative for traditional wastewater treatment technology (Vymazal 2001, 2007). These systems are robust, have a low cost of construction and maintenance and low external energy requirements, and are easy to operate and maintain, which makes them suitable for remediation of wastewater contaminants (Stottmeister et al. 2003; Vymazal 2010; Chandra and Kumar 2015b). The first experiments with the use of wetland plants for wastewater treatment were carried out in Germany in the 1950s (e.g., Seidel 1961), but full-scale systems were built only during the late 1960s (De Jong 1976). At the early stage of CW development, the application of CWs was mainly for the treatment of traditional tertiary and secondary domestic/municipal wastewater. However, from the last few decades, the application of CWs has also been significantly expanded to purify agricultural effluents, tile drainage waters, acid mine drainage, petrochemicals, food wastes, distilleries, pulp and paper mining, industrial effluents, landfill leachate, and urban and stormwater runoff throughout the world (Vymazal 2011). A number of different design variations are employed, which can be broadly categorized into two groups: (i) surface flow, where water is seen flowing on the surface of the water treatment system, and (ii) subsurface flow, where water flows through a porous medium such as sand or gravel. Direction of flow can be categorized as vertical or horizontal, and several operational regimes or modifications are possible as well (e.g., tidal flow regimes or added aeration). All of the different designs can be planted with various types of vegetation (Vymazal 2010, 2011). A schematic diagram of a typical CW is illustrated in Figure 7.1.

The wetland plants growing in CWs are important since they possess several functions in relation to water treatment (Brix 1997). The most commonly used plants are robust and fast-growing species of emergent wetland plants such as *Typha latifolia* and *Phragmites australis* (Vymazal 2001). The removal of numerous organic and

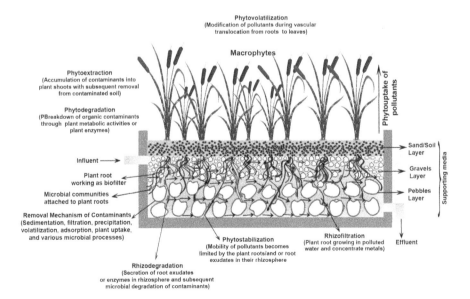

FIGURE 7.1 Schematic diagram showing a view of the developed constructed wetlands for the treatment of wastewater.

inorganic contaminants in CWs is complex and depends on a variety of removal mechanisms, including sedimentation, filtration, precipitation, volatilization, adsorption, adsorption–fixation reactions, and biological processes like plant uptake and microbial metabolic activities (Kadlec and Knight 1996; Kadlec et al. 2000). The CW system contains natural processes of aquatic macrophytes, the main biological component of wetland ecosystems that not only accumulate pollutants directly in their tissues but also act as catalysts for purification reactions that usually occur in rhizosphere of plants. Macrophytes indirectly facilitate the growth of important pollutant-reducing microorganisms through complex interactions in the rhizosphere (Brix 1994, 1997). In the rhizosphere, physico-chemical and biological processes are induced by the interaction of plants, microorganisms, and soil/sediments to remove heavy metals (HMs) and organic pollutants from wastewater (Brix 1997; Stottmeister et al. 2003). Hence, CWs are considered complex bioreactors in which different physico-chemical and biological processes with microbial communities, emergent plants, soil, and sediments accumulated in the lower layer take place in the systems. The complex microbial communities associated with wetland plant roots, created by interactions with effluent, are mainly responsible for the biodegradation efficiency of organic pollutants and ecosystem stability (Stottmeister et al. 2003; Kumar and Chandra 2018a).

7.3.1.1 Treatment of Wastewater in Constructed Wetlands Augmented with Rhizospheric/Endophytic Bacteria

Macrophytes perform effective phytoremediation on their own and provide a large surface area for microbial proliferation as well (Calheiros et al. 2009). The combined use of plants and microorganisms is a promising approach for the remediation of industrial and domestic effluents (Calheiros et al. 2007). In plant–bacteria synergism, plants support a microbial population, and in return, microbes improve plant growth and pollutant detoxification in wetland ecosystems. Various studies have been successfully carried out the world over by researchers to explore the role of augmented CWs in wastewater treatment. For instance, treatment of dye-containing wastewater by a developed lab-scale phytoreactor and enhancement of its efficacy by bacterial augmentation was investigated by Khandare et al. (2013). In this study, the authors used *Portulaca grandiflora* (Moss-Rose) to make a lab-scale phytoreactor. The efficacy of this phytoreactor was increased by augmenting *Pseudomonas putida* cultures to the soil. Four independent reactors with soil, bacteria, plants, and consortium were developed and used for the treatment of real textile effluent and a dye mixture. The textile effluent treated by the reactors containing soil, bacteria, and plants showed American Dye Manufacturers Institute (ADMI) removal by 31%, 50%, and 75% within 72 h, respectively, whereas the consortium reactor showed 89% color removal within 48 h. Effluent treated by the soil, bacteria, and plant reactors showed chemical oxygen demand (COD) reductions by 35%, 47%, and 59%; biological oxygen demand (BOD) by 14%, 28%, and 38%; total organic carbon (TOC) by 15%, 23%, and 37%; turbidity by 20%, 29%, and 41%; total dissolved solid (TDS) by 60%, 66%, and 71%; and total suspended solid (TSS) by 41%, 46%, and 60%, respectively, within 72 h, whereas the consortium reactor treated the effluent more efficiently, where COD, BOD, TOC, turbidity, TDS, and TSS of the effluent after treatment were reduced by 73, 54, 52%, 57%, 83%, and 71%, respectively, within just 48 h.

A plant–bacteria system within CWs to study the efficient remediation of tannery effluent (TE) was developed by Ashraf et al. (2018a). In a vertical-flow CW vegetated with *Leptochloa fusca* (Kallar grass), a consortium of three different endophytic bacteria, *Pantoea stewartii* ASI11, *Microbacterium arborescens* HU33, and *Enterobacter* sp. HU38, was used for bioaugmentation. CWs vegetated with *L. fusca* had the potential to remediate TE, but augmentation with endophytic bacteria enhanced the growth of *L. fusca* while aiding in the removal of both organic and inorganic pollutants from the TE. Moreover, the bacterial augmentation decreased toxicity in the TE as well.

Bioaugmentation is an effective treatment for pollutant removal from rural domestic wastewater (RDW) in CWs. Two subsurface flow CWs planted with *Typha orientalis* (CCW) and *Phragmites* (RCW) were constructed to study the effect of the bioaugmentation of *Paenibacillus* sp. XP1 on nitrogen removal from RDW in autumn (15–21°C) (Shao et al. 2013). CCW inoculated by *Paenibacillus* sp. XP1 (CCW-XP1) had obvious improvement on ammonia nitrogen (NH_3-N) and total nitrogen (TN) removal efficiency over RCW-XP1. The removal efficiency of TN in the CCW was similar to that of NH_3-N, and the maximal removal efficiency of 78% was achieved, double that of the control group. The final removal efficiencies of CCW-XP1 were found to be 73% for COD, 94% for NH_3-N, and 78% for TN. In order to determine whether bioaugmentation is an effective technique in CWs before the plants were harvested, N removal from a CWs planted with *Phragmites* was evaluated after inoculating with *Paenibacillus* sp. XP1 (Pei et al. 2016). The experiment was loaded with secondary effluent of RDW using the batch-loading method over a 17-day period in summer and autumn. COD, NH3-N, and TN decreased significantly in the CWs with *Phragmites* inoculated with *Paenibacillus* sp. XP1. Four days after treatments were set up, the removal efficiencies were found to be 76.2% for COD, 83% for NH_3-N, and 63.8% for TN in summer and 69.5% for COD, 76.9% for NH_3-N, and 55.6% for TN in autumn, which were higher than the control group without inoculation during the entire 17-day experiment. The inoculated bacteria did not have a noticeable effect on total TP removal in autumn. However, bioaugmentation still kept a low P concentration in the whole CWs. A summary of previous research works on the removal of organic matter, nutrients, and potentially toxic metals by CWs is described in Table 7.1.

7.3.2 FLOATING TREATMENT WETLANDS: A FLOATING TECHNOLOGY FOR WASTEWATER TREATMENT

FTWs are innovative phytoremediation systems for the treatment of wastewater that are simple in design and require no additional land, resulting in savings compared to conventional CWs. FTWs carrying different types of plants provide a sustainable solution for the remediation of wastewater due to their low cost and energy requirements (Ijaz et al. 2015; Afzal et al. 2019). FTWs can be natural or artificial, but, in either case, they combine the properties of natural ponds and

TABLE 7.1

Removal of Organic Matter, Nutrients, and Heavy Metals from Wastewater by Constructed Wetlands and Floating Treatment Wetlands Augmented with Rhizospheric and/or Endophytic Bacteria

Type of Wastewater	Name of Host Plant	Augmented Microorganisms	Pollutants Reduction	Technology Used	Study Level	Reference
Phenol-containing wastewater	*Phragmites australis*	*Acinetobacter lwoffii* ACRH76, *Bacillus cereus* LORH97, *Pseudomonas* sp. LCRH90	Reduction in COD, BOD, TOC	FTWs	Pilot scale	Saleem et al. (2019)
Tannery wastewater	*Leptochloa fusca*	*Pantoea stewartii* ASI11, *Microbacterium arborescens* HU33, *Enterobacter* sp. HU38,	Reduction in TDS, COD, BOD, TOC, NO_3^-, SO_4^-, Cl^-, Cr	CWs	Pilot scale	Ashraf et al. (2018a)
Oil field wastewater	*Brachiaria mutica*, *Phragmites australis*	*Bacillus subtilis* LORI66, *Klebsiella* sp. LCRI87, *Acinetobacter junii* TYRH47, *Acinetobacter* sp. LCRH81	Reduction in oil content (97%), COD (93%), BOD (97%)	FTWs	Field scale	Rehman et al. (2018)
Sewage effluent	*Brachiaria mutica*	*Acinetobacter* sp. BRSI56, *Bacillus cereus* BRSI57, *Bacillus licheniformis* BRSI58	Removal of organic and inorganic contaminants	FTWs	Pilot scale	Ijaz et al. (2015)
Oil-contaminated water	*Phragmites australis*, *Typha domingensis*, *Leptochloa fusca*, *Brachiaria mutica*	*Ochrobactrum intermedium* R2, *Microbacterium oryzae* R4, *Pseudomonas aeruginosa* R25, *Pseudomonas aeruginosa*, *Acinetobacter* sp. LCRH81, *Klebsiella* sp. LCRI-87, *Acinetobacter* sp. BRSI56, *P. aeruginosa* BRRI54, *Bacillus subtilis* LORI66, *Acinetobacter junii* TYRH47	Reduction in COD (97.4%), BOD (98.9%), TDS (82.4%), hydrocarbon content (99.1%), HMs (80%)	FTWs	Field scale	Afzal et al. (2019)

(Continued)

TABLE 7.1 (Continued)

Removal of Organic Matter, Nutrients, and Heavy Metals from Wastewater by Constructed Wetlands and Floating Treatment Wetlands Augmented with Rhizospheric and/or Endophytic Bacteria

Type of Wastewater	Name of Host Plant	Augmented Microorganisms	Pollutants Reduction	Technology Used	Study Level	Reference
Textile wastewater	*Phragmites australis, Typha domingensis*	*Acinetobacter junii, Pseudomonas indoloxydans, Rhodococcus* sp.	Reduction in color (97%), COD (87%), BOD (92%), HMs (87%–99%)	FTWs	Pilot scale	Tara et al. (2019)
Polluted river water and domestic wastewater	*Phragmites australis*	Five *Pseudomonas* sp. strains and one *Paenibacillus* sp. strain	Reduction in BOD, COD, NH_3-N, TP	CWs	Lab scale	Shao et al. (2014)
Rural domestic wastewater	*Typha orientalis, Phragmites*	*Paenibacillus* sp. XP1	Reduction in COD (73%), NH_3-N (94%), TN (78%)	CWs	Pilot scale	Shao et al. (2013)
Domestic wastewater	*Phragmites*	*Paenibacillus* sp. XP1	COD (76.2%), NH_3-N (83%), TN (63.8%) in summer and 69.5% for COD (69.5%), NH_3-N (76.9%), TN (55.6%) in autumn	CWs	Pilot scale	Pei et al. (2016)
Textile wastewater	*Brachiaria mutica*	*Microbacterium arborescens* TYSI04, *Bacillus endophyticus* PISI25, *Bacillus pumilus* PIRI30, *Bacillus* sp. TYSI1	COD (81%), BOD (72%), TDS (32%), color (74%), N (84%), P (79%), and HMs [Cr (97%), Fe (89%), Ni (88%), Cd (72%)]	CWs	Pilot scale	Hussain et al. (2018)
Textile wastewater	*Portulaca grandiflora*	*Pseudomonas putida*	COD (73%), BOD (54%), TOC (52%), turbidity (57%), TDS (83%), TSS (71%)	CWs	Pilot scale	Khandare et al. (2013)
Textile wastewater	*Glandularia pulchella*	*Pseudomonas monteilii* ANK	Reduction in ADMI, TOC, COD, BOD	CWs	Lab scale	Kabra et al. (2013)

Abbreviations: COD, Chemical oxygen demand; BOD, Biochemical oxygen demand; TOC, Total organic carbon; TDS, Total dissolved solids; TSS, Total soluble solids; P, Phosphorus; HMs, Heavy metals; N, Nitrogen; NH_3-N, Ammonia nitrogen; TP, Total phosphorous; TN, Total nitrogen; Cl⁻, Chloride; NO_3^-, Nitrite; ADMI, American Dye Manufacturers Institute; CWs, Constructed wetlands; FTWs, Floating treatment wetlands.

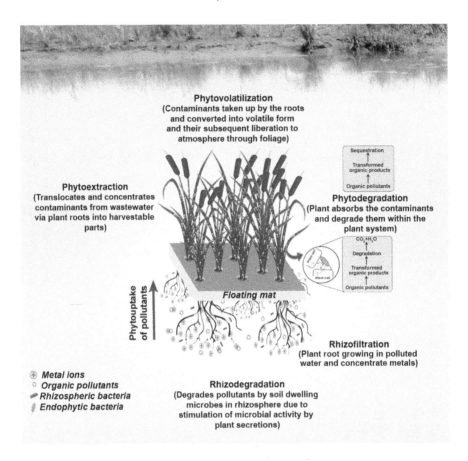

FIGURE 7.2 Schematic representation of floating treatment wetlands (FTWs) elucidating pollutant removal by plant–microbe mechanisms. (Modified from Saleem H et al. 2019. *Saudi Journal of Biological Sciences* 26(6):1179–1189.)

hydroponic floating vegetation (Chen et al. 2016). A schematic representation of a floating mat with holes for plantation of healthy seedlings of aquatic macrophyte and their mechanisms for remediation of wastewater pollutants is illustrated in Figure 7.2. FTWs harbor rooted and/or emergent plants of aquatic or terrestrial origin, grown on self-buoyant mats being fixed on wastewater. Plant roots hanging down into the water column not only act as a natural filter for pollutant removal but also provide a large biological surface area for enhanced growth of microorganisms and development of biofilm formation for the physical entrapment, biochemical transformation, and degradation of organic pollutants (Brix 1994; Zhang et al. 2014). In addition, the plant roots not only provide for attachment of microorganisms, but plants also provide a habitat for wildlife. The success of such a system purely relies on the microbial community in the biofilm as well as on the plant species (Weyens et al. 2009a; Ashraf et al. 2018a,b). Phytoremediation is a vital natural process of pollutant removal from aquatic ecosystems. Industrial effluent is rich in toxic compounds that can inhibit plant growth and microbial proliferation, consequently affecting remediation efficiency. To reverse the effect of these toxic compounds, the combined application of plants and contaminant-degrading bacteria has recently been proposed (Khandare et al. 2013; Arslan et al. 2017).

7.3.2.1 Treatment of Wastewater in Floating Treatment Wetland Augmented with Rhizospheric/Endophytic Bacteria

Plants, in combination with bacteria, have been shown to be more effective for biotransformation and complete biodegradation of organic and inorganic loaded effluents. Plants provide residency and nutrients to the rhizo- and endophytic bacteria, whereas bacteria help their hosts gain more biomass by reducing contaminant stress and performing plant growth-promoting services (Khan et al. 2014). Further, the efficacy of FTWs can be enhanced by inoculation with bacteria for wastewater applications of, for example, municipal as well as domestic and industrial origins. Rehman et al. (2018) investigated the potential of plant–bacteria synergism in FTWs for the efficient remediation of oil field wastewater. Two plants, *Brachiaria mutica* and *Phragmites australis*, were vegetated on floatable mats to develop FTWs and inoculated with a developed bacterial consortium, which were then inoculated with a consortium of the hydrocarbon-degrading bacteria *Bacillus subtilis* LORI66, *Klebsiella* sp. LCRI87, *Acinetobacter junii* TYRH47, and *Acinetobacter* sp. LCRH81. Both plants successfully removed organic and inorganic pollutants from wastewater, but bioaugmentation of *P. australis* significantly enhanced the plant's efficiency to reduce oil content (97%), COD (93%), and BOD (97%) in wastewater. Analysis of alkane-degrading gene (*alkB*) abundance and its expression profile further validated higher microbial growth and degradation activity in the water around *P. australis* as well as its roots and shoots. This study provides insight into the available phytotechnology for remediation of crude oil-contaminated water and introduces a wetland macrophyte, *P. australis*, with a tailor-made bacterial consortium as an effective tool for improved phytoremediation efficiency of FTWs.

A terrestrial plant, *B. mutica*, was also used with inoculation of three endophytic bacterial strains, *Acinetobacter* sp. BRSI56, *Bacillus cereus* BRSI57, and *Bacillus licheniformis* BRSI58, to develop FTWs for the remediation of sewage effluent (Ijaz et al. 2015). Results indicated that *B. mutica* has the potential to remove both organic and inorganic contaminants from sewage effluent. However, endophytic inoculation in FTWs further enhanced the removal efficiency. Maximum reduction in COD, BOD, TN, and phosphate (PO_4^{2-}) was achieved by the combined use of plants and bacteria.

FTWs can be used for the treatment of textile effluent, and their working efficiency can be improved by plant–bacterial synergism. Tara et al. (2019) evaluated the effects of bacterial augmentation on the efficiency of FTWs to remediate textile wastewater. Two wetland plants, *P. australis* and *Typha domingensis,* were used to develop FTWs, which were then augmented with a bacterial consortium of three strains (*Acinetobacter junii, Pseudomonas indoloxydans*, and *Rhodococcus* sp.). Both plant species removed color, organic matter, toxicity, and HMs from textile wastewater, and their removal efficiency was further enhanced by augmentation with bacteria. The maximum removal efficiencies of color, COD, and BOD after an 8-day period were

97%, 87%, and 92%, respectively, by FTWs carrying *P. australis* inoculated with the bacterial consortium. Furthermore, the same combination showed a 87%–99% reduction of HMs in the textile wastewater as well. A summary of previous research works on the removal of organic matter, nutrients, and potentially toxic metals by FTWs is described in Table 7.1.

8 Arbuscular Mycorrhizal Fungi-Assisted Phytoremediation of Organic and Inorganic Pollutants

8.1 INTRODUCTION

Soil and water pollution with toxic heavy metals (HMs) and organic pollutants due to rapid industrial activities poses a serious environmental and human health problem (Tchounwou et al. 2012; Chandra et al. 2018a,b). Therefore, soil clean-up and recultivation has become a priority of environmental engineers and biotechnologists. Over the last decades, numerous physico-chemical methods/approaches including solvent extraction techniques, chemical oxidation, soil stabilization and solidification, and thermal treatments developed for soil remediation have usually been highly efficient; however, they are expensive, labor intensive, and energy consuming and change the soil structure, decrease soil microbial activity, and lead to the depletion of the nutrients that are essential for plant vegetation (Liu et al. 2018). Besides, the choice of any method would depend on the type of pollutant to be eliminated, the type of use that would be made of the polluted site after its rehabilitation, and available time and finances (Selvi et al. 2019). The most conventional soil remediation methods are effective in small areas, but they need special equipment (Liu et al. 2018). In contrast, phytoremediation has been regarded as an effective, nonintrusive, inexpensive, promising biological method proposed for cleanup of contaminated environments (Sessitsch et al. 2013; Thijs et al. 2017). This technology involves plants and their associated microorganisms to remove toxic inorganic and organic pollutants through their metabolic activities (Glick 2003; Rajkumar et al. 2012). It has been demonstrated that involvement of the catabolic potential of both microorganisms and plants is the most effective approach to take up and degrade soil contaminants (Ma et al. 2009a,b). In this context, the use of arbuscular mycorrhizal fungi (AMF) and their role in phytoremediation has emerged as a new and interesting choice to clean up soil and water contaminated with organic and inorganic pollutants (Alarcon et al. 2008; Rajtor and Piotrowska-Seget 2016). Arbuscular mycorrhizal associations are integral, functioning parts of plant roots; are widely recognized as enhancing phytoremediation efficiency of environmental pollutants; and are often exploited to improve this because of their plant growth-promoting, pollutant-degrading, and/or detoxification capacities (Chen et al. 2018a,b; Begum et al. 2019). They are reported to be present on the roots of plants growing on metalliferous soils and play an important role in metal tolerance and accumulation (Yang et al. 2016). About 20 000 plant species are unable to grow or survive without microbial symbiotic interactions. Therefore, the development of every biotechnological application involving plants and plant-associated microbes should be taken into consideration.

8.2   ECOLOGY OF ARBUSCULAR MYCORRHIZAL FUNGI

AMF are nonpathogenic soil-borne microorganisms that form a mutualistic symbiotic association with the root system of about 80%–85% of terrestrial plants, providing a direct physical link between soil and plant root. Arbuscular mycorrhiza symbiosis is thought to be a largely promiscuous association between >100 000 plant species and a few 100 AMF morphotypes, which have long been regarded as the equivalent of species (Chen et al. 2018a,b). This symbiosis is increasingly being recognized as an integral and important part of natural ecosystems throughout the word. The fungus colonizes the root cortex and develops a dense external mycelium around the root. This mycelium acts as an intermediary between the soil and plant, absorbing nutrients from the soil and transporting them to the host root (Begum et al. 2019). In turn, fungi obtain photosynthates from the plant for their metabolic functions. As obligate biotrophs, these fungi are unable to complete their life cycle in the absence of the host plant.

AMF are ubiquitous rhizosphere microbes occurring in almost all habitats and climates, including metal-contaminated soils, and facilitate host plants to uptake and transport phosphorus (P) and other relatively immobile soil nutrients, promote plant growth, and enhance their stress tolerance (Gaur and Adholeya 2004). Many reports have shown that AMF may play an important role in soil phytoremediation processes (Meier et al. 2012; Yang et al. 2015; Elhindi et al. 2018). AMF are obligatory plant root endosymbionts that were formerly included in the order Glomales in the phylum Zygomycota, but they have recently been moved to a new phylum, Glomeromycota (Meier et al. 2012). AMF of Glomeromycota are the most common soil microorganisms in natural and agricultural soils. Based on morphological and molecular identification methods, AMF were divided into 12 genera (*Acaulospora, Ambispora, Archaeospora, Diversispora, Entrophospora, Kuklospora, Geosiphon, Gigaspora, Glomus, Intraspora, Paraglomus, Scutellospora*), with the *Glomus* species dominant. However, due to the relatively few distinctive morphological features of AMF (primarily associated with spores), and due to their essentially asexual mode of propagation, the traditional species concept is problematic in the context of AMF.

The establishment of AMF symbioses begins with the colonization of a compatible root by hyphae produced by AMF soil propagules, asexual spores, or mycorrhizal roots. This is followed by appressorium formation and enters into the root cortex to form specialized finely branched hyphae, called the arbuscules. The surface area of the hyphae is approximately 100 times greater than the surface area of the root, and their length can be several orders of magnitude greater than that of the plant root system; therefore, they can explore a larger volume of soil than the roots alone. Such an extensive network of fungal mycelium helps to release nutrients and organic contaminants from soil particles and thus facilitate nutrient and water uptake by plants (Meier et al. 2012; Cabral et al. 2015). Soil microorganisms are known to produce compounds that increase the rates of root exudation. This, in turn, stimulates AMF mycelia in the rhizosphere or facilitates root penetration by the fungus. Before colonization, it is assumed that a continuous dialogue of plant-derived signals is exchanged between the symbionts to establish colonization (Hodge 2014). Plant roots release several organic compounds like sugars and amino acids in form of root exudate, which are potential fungal stimuli, but phenolic compounds, particularly flavonoids, are known as key signaling components in many plant–fungal interactions (Hodge 2014; Begum et al. 2019). AMF hyphae create a widespread underground network that constitutes a direct connection between plant roots, soil, and associated microflora. Entanglement of the AMF extraradical mycelium is responsible for the maintenance of soil structure and improves soil water-holding capacity (Hata et al. 2010). It has been demonstrated that AMF exudates directly impact soil bacterial community composition, and some bacteria associated with AM can improve colonization, root branching, and antifungal properties (Bonfante and Genre 2010). The establishment of the AM fungus in the root cortex is known to change many key aspects of plant physiology. These include the mineral nutrient composition of plant tissues, the hormonal balance, and the patterns of C allocation. Therefore, AM symbiotic status changes the chemical composition of root exudates, while the development of an AM soil mycelium, which can act as a carbon source for microbial communities, introduces physical modifications into the environment surrounding the roots (Bonfante and Genre 2010; Meier et al. 2012; Cabral et al. 2015). AM-induced changes in plant physiology affect microbial populations, both quantitatively and qualitatively, in either the rhizosphere and/or the rhizoplane. Therefore, the rhizosphere of a mycorrhizal plant can have features that differ from those of a non-mycorrhizal plant.

8.3 ARBUSCULAR MYCORRHIZAL FUNGI-ASSISTED PHYTOREMEDIATION

Phytoremediation, an emerging technique that uses plants to clean up soil and water, has gained technical attention and public acceptance thanks to its low cost and minimal impact on soil properties (Salt et al. 1995; Dickinson et al. 2009). Depending on the type of pollutant, different strategies for phytoremediation, such as phytostabilization, phytodegradation, phytoextraction, and rhizodegradation, can be used. (Ashraf et al. 2019) Moreover, several approaches have been explored in recent decades to boost the efficiency of this technology, such as microbe-assisted phytoremediation, in particular by AMF (Schneider et al. 2016). AMF are known for their beneficial effects on plant growth and development and their strong degradative and detoxification capacities and are therefore also often applied during phytoremediation of environmental pollutants (Cabral et al. 2015). AMF increases soil mineral uptake, including HMs, and contributes to phytoprotection by enhancing nutritional and physiological processes in plants. Such processes include metal immobilization in roots and fungal mycelium, which may result in reduced metal accumulation in plant roots (Meier et al. 2012). Another phytoprotective effect mediated by AMF is the complexation of chemical elements with glomalin molecules, an iron-containing, heat-stable glycoprotein, which are exuded in the soil by these fungi, possibly leading to lower bioavailability of heavy metals and increased plant growth (Rillig et al. 2002; González-Chávez et al. 2004). Glomalin has been operationally defined as glomalin-related soil protein (GRSP) by extraction and detection conditions from soil, and it is detected in large amounts in diverse ecosystems (Yang et al. 2017). The sticky GRSP acts as biological glue, helping to bind soil tiny particles into small aggregates of different sizes. Well-aggregated soil is stable enough to resist wind and water erosion and has better air and water infiltration rates favorable for plant and microbial growth. Additionally, GRSP is recalcitrant enough to have a long residence time in soils and plays a pivotal role in long-term carbon/nitrogen storage and HM sequestration (Jia et al. 2016). Therefore, the release and accumulation of GRSP in soils can be a very important mechanism for ecological restoration of heavy metal–contaminated soils (Chern et al. 2007). The establishment of AMF in plant roots has been shown to reduce damage caused by soil-borne plant pathogens with an enhancement of plant resistance/tolerance in mycorrhizal plants. In any case, the effectiveness of AM in biocontrol is dependent on the AMF involved, as well as the substrate and the host plant. Besides heavy metal phytoremediation, AMF are able to facilitate plant establishment and survival in organic pollutant–contaminated soil by protecting plants against the phytotoxicity of organic pollutants but also enhance soil bioremediation by stimulating telluric microbial activity and improving soil structure. Indeed, plants may take organic contaminants into their tissue and transform or mineralize the contaminants (Schneider et al. 2016). They may also exude oxidative enzymes that contribute to the degradation of organic pollutants or, through exudation, may provide easily degradable carbon. The latter provides energy that can increase microbial activity in soil, which may enhance the degradation of organic pollutants through direct metabolism or co-metabolism. In this context, a diverse and functional microbial population may play an important role in plant survival and growth.

8.3.1 Remediation of Heavy Metals

AMF have great potential for assisting HM hyperaccumulators in the remediation of contaminated soils (Miransari et al. 2011). The combined use of plants with AMF has advantages over the use of hyperaccumulators alone and has been proposed as one of the most promising green remediation techniques (Yang et al. 2016). Hyperaccumulating plants are usually nonmycorrhizal with a slow growth rate and produce little biomass, and long periods of time are required for cleaning up HMs in soil, but there are several reports of the presence of AM in hyperaccumulating plants (Chandra et al. 2018a,b). These plants are capable of accumulating high concentrations of HMs under natural conditions and produce biomass that exceeds most other hyperaccumulators. AMF are important regulators of plant performance in soils contaminated with HMs and increase the resistance of plants to HM toxicity (Abbaslou et al. 2018). In addition, it has been demonstrated that AMF can increase the HM translocation factor, biomass, and trace element concentrations of plants (Schneider et al. 2016;

Singh et al. 2019). Mycorrhizal plants also improve phytostabilization because HMs (Zn, Cd, Cu) are confined to hyphae and roots without translocating these elements to aerial parts (Joner and Leyval 1997, 2001). As such, the metals remain in the soil, but because they are less bioavailable, toxicity to other organisms is reduced (Leyval et al. 2002). For phytoremediation of soil polluted with HMs, the phytostabilization strategy involves the immobilization of HMs in the soil by establishing plants. This reduces both soil erosion and transfer of the HMs to aquifers, thus avoiding their dispersion by the wind (Chen et al. 2018a,b). Alternatively, phytoextraction takes advantage of the ability of plants to hyperaccumulate metals. During greenhouse studies with sunflowers grown in soils contaminated with three different Cd concentrations, successful AMF colonization by *Rhizophagus irregularis* resulted in enhanced phytoextraction of Cd, whereas *Funneliformis mosseae* enhanced phytostabilization of Cd and Zn (Hassan et al. 2013). These results indicated that the two roots associated with AMF may adopt several mechanisms to trigger either metal mobilization or immobilization, therefore contributing directly to phytoextraction or phytostabilization processes. HM phytoremediation with *Rosmarinus officinalis* assisting AFM (*Glomus mosseae* and *Glomus intraradices*) and fibrous clay minerals (incubated with 8% and 16% clay) were performed in a greenhouse on contaminated soils (Abbaslou et al. 2018). Toxic levels of Cd, Pb, Zn, and Cu in soil decreased after cultivation to a less-than-critical threshold. Accumulation of elements in roots and shoots had the sequence Cu>Zn>Mn>Cd>Pb>Fe. Soils were incubated with fibrous clay minerals to remediate them. Soils adsorbed heavy metals in the sequence Pb>Cd>Zn = Fe>Cu>Mn. In general, rosemary with moderate salinity and aridity tolerance could adsorb toxic elements through phytostabilization and phytoextraction, mostly phytoextraction. Bioaugmentation-assisted phytoremediation and immobilization of elements by fibrous minerals enhanced remediation of soils by promoting plant growth and retention of elements. Although AM fungi do not necessarily stimulate phytoextraction, the potential to increase the biomass of the plants, to enhance nutrient and water uptake, and to improve soil conditions are important reasons to include AM fungi in further research (Turnau et al. 2005). Examples of AMF application for improvement of the efficiency of heavy metal phytoremediation are listed in Table 8.1.

8.3.2 REMEDIATION OF ORGANIC POLLUTANTS

AMF can play a role in two aspects of organic pollutant phytoremediation (Li et al. 2006). They can not only improve plant establishment and growth on soil polluted with organic compounds but also enhance the dissipation, degradation, or mineralization of persistent organic pollutants (POPs) (Lenoir et al. 2016; Fecih and Baoune 2019). Degradation is the process by which organic substances are broken down into smaller compounds. When degradation is complete, the process is called mineralization. In contrast, dissipation is the sum of all processes involved in the disappearance of the organic substance, such as the uptake of the organic pollutant into plants, the adsorption onto roots, and degradation by the microbial community (Lenoir et al. 2016). In general, dissipation is used when the process of the disappearance is unknown. Joner and Leyval (2003) conducted a time-course pot

TABLE 8.1

Examples of Arbuscular Mycorrhizal Fungi Application for Improvement of the Phytoremediation Efficiency of Heavy Metal–Contaminated Soil

Name of Fungi	Host Plant	HM-Contaminated Soil	Effect on Plant/Soil	Mechanisms	Reference
Rhizophagus clarus, *Acaulospora colombiana*	*Enterolobium contortisiliquum* (Vell.) Morong (pacara earpod tree)	Coal-mining waste Contaminated soil	Soil glomalin concentrations (2.98 mg kg^{-1}), phytoprotection	Phytoextractive	Leandro dos Santos et al. (2017)
Rhizophagus intraradices	*Robinia pseudoacacia*	Pb	Influencing plant biomass, plant photosynthesis, macronutrient acquisition, N, P, S, Mg	Phytoremediation	Yang et al. (2016)
Funneliformis geosporum	*Triticum aestivum* L. cv. Gemmeza-10	Zn	Dry weight increased, translocation of Cu and Zn concentrations in the root, shoot, and grain	P, S, K, Ca, and Fe concentration was decreased	Abu-Elsaoud et al. (2017)
Arbuscular mycorrhizal fungi	*Sophora viciifolia*	Pb, Zn	Soil organic matter and organic carbon subsequently influencing the aggregate formation and particle-size distribution in HM-polluted soils	Synthesis of glomalin	Yang et al. (2017)
Boehmeria nivea	*Glomus, Acaulospora, Funneliformis, Claroideoglomus, Gigaspora, Paraglomus, Rhizophagus*	Sb	Colonization increased consistently with the increasing Sb concentrations in the soil	Bioremediation	Wei et al. (2015)

(Continued)

TABLE 8.1 (*Continued*)

Examples of Arbuscular Mycorrhizal Fungi Application for Improvement of the Phytoremediation Efficiency of Heavy Metal–Contaminated Soil

Name of Fungi	Host Plant	HM-Contaminated Soil	Effect on Plant/Soil	Mechanisms	Reference
G. intraradices	Populus generosa, Salix viminalis	Cd, Cu, Pb, Zn	Cu and Cd translocation to the shoots was limited and presented higher levels of Cu in the roots	Exclusion mechanism	Bissonnette et al. (2010)
Elsholtzia splendens	Gigaspora margarita, G. decipiens, Scutellospora gilmori, Acaulospora spp., Glomus spp.	Cd, Cu, Pb, Zn	Increased the concentration and accumulation of P, Cu, Zn, Pb, and Cd in plants	Phytoextraction	Wang et al. (2005, 2007)
Tree species (Searsia lancea, S. pendulina, Tamarix usneoides)	Acaulospora, Claroideoglomus, Diversispora, Glomus, Sclerocystis	Cr, Ni, Zn, Pb	All promising candidates for phytoremediation programs	–	Spruyt et al. (2014)
Silene vulgaris, Thlaspi caerulescens, Zea mays	Glomus mosseae Glomus sp., Glomus constrictum	Metal-contaminated landfill	Sulfur supplement increased vesicular root colonization	Phytoextraction	Pawlowska et al. (2000)
Rosmarinus officinalis	Glomus mosseae, Glomus intraradices	Cu, Zn, Mn, Cd, Pb, Fe	Accumulation of Cu, Zn, Mn, Cd, Pb, Fe in roots and shoots; decreased toxic levels in soil	Phytostabilization and phytoextraction	Abbaslou et al. (2018)
Zea mays L.	Glomus caledonium	Zn	Zn translocation from roots to shoots was enhanced	Phytoextraction	Chen et al. (2004)
Nicotiana rustica L. var. Azteca	Glomus intraradices Schenck & Smith	Zn	Zn concentration increased in tested plant	Immobilisation	Audet and Charest (2006)
Pteris vittata L.	Glomus mosseae, Glomus, Glomus caledoniumintraradices	As, U	Mycorrhizal colonization significantly increased root U concentrations	Phytoextraction	Chen et al. (2006a)
Pteris vittata, Cynodon dactylon	Glomus mosseae, Glomus intraradices	As	Tested plants accumulated higher N and P		Leung et al. (2013)

experiment to measure the dissipation of polycyclic aromatic hydrocarbons (PAHs) in the rhizosphere of clover and ryegrass grown together on two industrially polluted soils containing 0.4 and 2 g kg^{-1} of 12 PAHs. The impact of the fungal root symbiosis AMF on PAH degradation was also assessed. The two soils behaved differently with respect to the time-course of PAH dissipation. The less polluted and more highly organic soil showed low initial PAH dissipation rates, with small positive effects of plants after 13 weeks. At the final harvest (26 weeks), the amounts of PAHs extracted from nonplanted pots were higher than the initial concentrations. In parallel planted pots, PAH concentrations decreased as a function of proximity to roots. The most polluted soil showed higher initial PAH dissipation (25% during 13 weeks), but at the final harvest, PAH concentrations had increased to values between the initial concentration and those at 13 weeks. An effect of root proximity was observed for the last harvest only. The presence of mycorrhiza generally enhanced plant growth and favored growth of clover at the expense of ryegrass. Mycorrhiza enhanced PAH dissipation when plant effects were observed. However, the efficiency of this method of remediation depends on the species and origin of the fungi used, the type of plant colonized, and the concentration of the contaminant. Successful studies showing major benefits of inoculated AMF for the phytoremediation of contaminated soils by organic pollutants are listed in Table 8.2.

8.3.3 MYCORRHIZATION HELPER BACTERIA

Microbial populations in the rhizosphere are known either to interfere with or to benefit the establishment of mycorrhizal symbioses. Mycorrhizal fungi and bacteria in the rhizosphere can interact with each other at different levels of cellular integration, ranging from apparently simple associations, through surface attachment, to intimate and obligatory symbiosis (Tarkka and Frey-Klett 2008; Rigamonte et al. 2010). A typical beneficial effect is that exerted by "mycorrhization helper bacteria" (MHB), a term that was coined by Garbayetansley (1994) for those bacteria known to stimulate mycelial growth of mycorrhizal fungi and/or enhance mycorrhizal formation. This synergism may not only be important in promoting plant growth and health but may also be significant to rhizosphere ecology (Kurth et al. 2013; Labbe et al. 2014). The possible mechanisms for the bacterial helper effects could directly influence the germination and growth rate of fungal structures, root development, and root susceptibility to fungal colonization, including production of phytohormones, amino acids, and/or cell wall hydrolytic enzymes (Khan et al. 2006). The majority of MHB that have been described so far belong to the fluorescent pseudomonads and sporulating bacilli. It seems that MHB include a variety of Gram-negative and Gram-positive species, suggesting that their activities could perhaps be found in all bacterial groups that exist in the rhizosphere (Tarkka et al. 2008). Mycorrhizae are often described as tripartite interactions, because in their natural environment, bacteria are associated with arbuscular and ectomycorrhizal fungi by colonizing the extraradical hyphae or as endophytic bacteria living in the cytoplasm of at least some fungal taxa. Several reports have demonstrated enhanced AMF colonization in roots in the presence of MHB (Rigamonte et al. 2010; Labbe et al. 2014; Jambon et al. 2018). Also, endobiotic bacteria, that is, *Candidatus* and *Glomeribacter gigasporarum*, residing

TABLE 8.2

Examples of AMF Application for Improvement of the Efficiency of Hydrocarbon Removal

Name of Fungi	Host Plant	Name of Site	Plant Growth-Promoting Effect	Reference
Glomus mosseae	*Triticum aestivum cv.* Sakha 8, *Vigna radiata* V 2010, *Solanum melongena* L.	Polycyclic aromatic hydrocarbons contaminated environment	Phytouptake, biodegradation	Root lipid, peroxidase, catalase, and polyphenol oxidase in mung bean, root lipid, and polyphenol oxidase in wheat plant	Rabie (2005)
Claroideoglomus, Diversispora, Rhizophagus, Paraglomus	*Eleocharis obtusa, Panicum capillare*	Petroleum hydrocarbon–polluted sedimentation basin	Bioremediation	—	de la Providencia et al. (2015)
Glomus mosseae fungal spores	*Lolium perenne* L., cv. Barclay, *Trifolium repens* L., cv. grasslands huia	Polycyclic aromatic hydrocarbonscontaminated soil	Enhanced organic pollutants dissipation	Enhanced plant growth and favored growth of clover at the expense of ryegrass	Joner and Leyval (2003)

Abbreviation: PAH, polycyclic aromatic hydrocarbons.

inside fungal hyphae can have positive effects on mycorrhizal growth (Vannini et al. 2016). Managing the microbial population in the rhizosphere by using an inoculum consisting of a consortium of PGPR, MHB, nitrogen-fixing rhizobacteria, and AMF as allied colonizers and biofertilizers could provide plants with benefits crucial for ecosystem restoration on derelict lands (Khan et al. 2005).

8.3.4 MECHANISMS OF HEAVY METAL ALLEVIATION BY ARBUSCULAR MYCORRHIZAL FUNGI DURING PHYTOREMEDIATION

AMF can also play a role in the protection of roots from metal(loid) toxicity by mediating the interactions between the metals and the plant roots. To ensure their survival in metal-contaminated soils, AMF can use different extracellular and intracellular defense mechanisms. Most of the mechanisms that plants and AMF adopt to alleviate metal stress are quite similar because of the strict biotrophy of AMF. Extracellular mechanisms, such as chelation and cell wall binding or biosorption of metals to biopolymers in the cell wall, such as chitin and glomalin, may be used to prevent metal uptake. The wall of fungal cells is mainly composed of polysaccharides and chitin, which can act as a barrier to metal ions and other solutes and control their uptake into the cell. The presence in the cell wall of free amino acids together with hydroxylic, carboxylic, and other functional groups confers a negative charge to the structure, allowing it to bind ionic elements, including most metals present in soil. It is known that the extra-radical mycelium (ERM) of AMF can contribute to the remediation of metal-polluted soils. Joner and Leyval (1997) and Joner et al. (2000) reported that the outer surface of ERM has a larger capacity for sorbing metals than root cells. In comparing the active sorption of Cd and Zn by several *Glomus* species, they found that the ERM in *Glomus mosseae* could sorb 0.5 mg Cd g^{-1}, which was 10 times higher than that observed with other fungal biosorbents such as *Rhizopus arrhizus* (Zhou 1999). Similarly, Gonzalez-Chavez et al. (2002, 2009) concluded that the ERM could sorb and accumulate high levels of Cu (3–14 mg Cu g^{-1} dry hyphae). They also described the existence of several tolerance mechanisms in different AMF isolated from the same Cu-polluted soil. Other mechanisms by which the ERM of AMF can immobilize metals include the production of extracellular glycoproteins. By reducing metal bioavailability and promoting phytostabilization, the ERM of AMF minimizes the exposure of plants to metals. The mechanisms used by AMF are summarized in Figure 8.1. Intracellular mechanisms, including binding to nonprotein thiols and transport into intracellular compartments, can reduce the concentration in the cytosol. Intracellular mechanisms depend on transporter proteins and intracellular chelation through metallothioneins (MTs), glutathione, organic acids, amino acids, and compound-specific chaperones. Metal transporter proteins can alleviate metal stress by subcellular compartmentation via transporters into the vacuole or other internal cell compartments and/or vacuolar compartmentation of a complex (e.g., the GSH-M complex). Once chelated, these metal complexes can be transported as well. Furthermore, uptake/efflux of metals via specific transporter systems located in the plasma membrane can be down regulated. Additionally, antioxidative defense processes to detoxify reactive oxygen species (ROS), such as the superoxide anion radical, hydrogen peroxide, hydroxyl radicals, and the single oxygen, and mechanisms

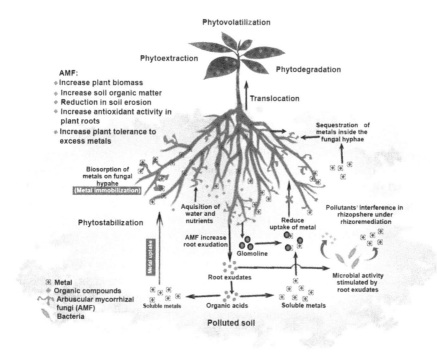

FIGURE 8.1 Mechanism showing the remediation of heavy metals and organic pollutants through plant–fungal and bacteria synergism.

that focus on the repair of metal-damaged biomolecules. The membrane transporters in AMF arbuscules may carry metals to the interfacial matrix (the contact zone between the plasma membrane of the fungus and the plant cell) and their subsequent incorporation inside the plant. This may explain how some mycorrhizal plants can accumulate metals in their shoots. It is also possible that the fungi assign some of their structures (spores) to store metals. The storage of metals in spores has been described only in monoxenic culture; however, recently, Meier (2011) demonstrated the accumulation of Cu in mycorrhizal spores in soils contaminated with high Cu concentration. These observations and the positive effect of mycorrhizal symbiosis on the phytoremediation of metal-polluted soils are of great biotechnological interest, as mycorrhizal plants are as effective in extracting metals (e.g., Cu, Cd, Pb, Zn) as nonmycorrhizal hyperaccumulator plants.

8.3.5 MECHANISMS OF ORGANIC POLLUTANT REMEDIATION BY ARBUSCULAR MYCORRHIZAL FUNGI DURING PHYTOREMEDIATION

Synergistic interactions between bacteria and fungi can not only promote plant growth and development, they can also stimulate the biodegradation of organic pollutants. The significant role of AMF in cleaning up organic pollutants, especially hydrocarbon-contaminated sites, is mainly based on AMF-mediated rhizodegradation. The ephemeral extraradical mycelium extends the range of the rhizosphere, thus creating the hyphosphere, which constitutes a favorable habitat for activity of soil microorganisms and their proliferation. Living fungal hyphae directly translocate photosynthetically derived carbon in the soil in the form of energy-rich, low-molecular-weight sugars and organic acids as well as unidentified high-molecular-weight polymeric compounds. In the processes of rhizodegradation, both enzymes of microbial origin and root exudates are involved. The latter, apart from plant extracellular enzymes, contain a range of co-metabolites, which constitute the source of carbon and energy for microorganisms responsible for biodegradation of organic contaminants. The co-metabolites that stimulate PAH degradation primarily include phenolic compounds, dissolved organic carbon, and low-molecular-weight organic acids. These telluric microorganisms possess enzymatic systems that catalyze degradation–detoxification reactions of organic pollutants by oxidation (i.e., cytochrome P450 monooxygenases, dioxygenases, peroxidases, laccases, and polyphenol oxidases), reduction (i.e., nitroreductases), hydrolysis (i.e., nitrilases and phosphatases), and conjugation (i.e., glutathione-S-transferases). These enzymes are mainly responsible for the oxidative transformation and degradation of PAHs and additionally protect plants from oxidative stress. It is has been found that the activity of plant enzymes may be modified due to mycorrhizal infection. Bacterial–fungal interactions can promote phytoremediation of organics by facilitating the access of degrading microorganisms to the pollutant. As mentioned earlier, fungal mycelia can, after all, serve as a highway for pollutant-degrading bacteria in water-unsaturated environments, leading to their dispersal in polluted soil. Moreover, bacteria and fungi can complement each other in the degradation pathway of the pollutant. This co-metabolic degradation can lead to full degradation of the pollutant, where one can further degrade the intermediates formed by the other. These mechanisms are described in detail subsequently and presented in Figure 8.1.

9 Genetically Engineered Microbes in Phytoremediation of Organic and Inorganic Pollutants

9.1 INTRODUCTION

The increasing contamination of soil, sediment, and water with heavy metals and organic compounds by natural and industrial processes is a worldwide problem (Chandra et al. 2018a,b). Therefore, the removal of unnecessary heavy metals and organic compounds from polluted soils has become an increasingly pressing issue. Traditional civil and engineering technologies are often too expensive for the remediation of the most polluted sites. Thus, the use of plants and their associated microorganisms to degrade and detoxify contaminants in the laboratory, as well as at the actual site of contamination, known as phytoremediation, appears an attractive option (Chandra et al. 2015; Kumar and Chandra 2018a). This technology has emerged as a cost effective, non-invasive, and publicly acceptable way to address the removal of environmental contaminants. Although phytoremediation is a promising option, it also has drawbacks. Pollutants above a certain level can be toxic to both the plants and the associated microorganisms (van Dillewijn et al. 2008), plant metabolism can transform the contaminant (at least temporarily) into a more toxic chemical, or the plant can mobilize the contaminant from the soil to an aerial part where it can be introduced into the food chain (Finkel et al. 2017). To overcome these issues, genetic engineering of symbiotic plant-associated microbes (i.e., plant growth-promoting rhizobacteria [PGPR] and endophytes) could also be a promising phytobacterial technology to enhance tolerance to high metal concentrations and detoxification degradation of toxic compounds and is considered one of the most promising new technologies for remediation of contaminated environmental sites (Sriprang et al. 2002; Barac et al. 2004; Germaine et al. 2006). It involves the introduction of one or more genes of interest that code for enzymes responsible for enhanced remediation, stress tolerance, metal chelators, uptake regulators, transporters, and homeostasis. Generally, plants and microbes coexist in nature in any given ecosystem, and they may be symbiotic with or compete with one another for their survival (Shim et al. 2000; Vangronsveld 2009; Weyens et al. 2009a,b,c,d). Bacteria that reside in different compartments of plants can synthesize several compounds that assist plants in overcoming stress, providing essential nutrients required for plant growth and development, improving plant defense system against pathogens, and stimulating contaminant degradation (Glick 2003, 2012; Newman and Reynol 2005; Wu et al. 2006a,b). Bacteria (rhizospheric and/or endophytic) can be engineered via natural gene transfer or recombinant DNA technology to produce specific enzymes capable of degrading toxic organic pollutants found in the environment. Symbiotic relationships between genetically engineered bacteria and plants have been widely exploited for *in situ* bioremediation of various organic pollutants. The aim of this chapter is to summarize recent knowledge on the degradation and detoxification of environmental pollutants through plant-microbe synergism and to critically discuss the potential and limitations of microbe-assisted phytoremediation approaches for the recovery of polluted ecosystems.

9.2 GENES INVOLVED IN ASSISTED PHYTOREMEDIATION OF ORGANIC AND INORGANIC POLLUTANTS

Improvement of contaminant removal by the introduction of degradation genes in endophytes and rhizospheric bacteria has also been demonstrated, not only as a way to increase the degradation rate of certain pollutants but also as a way to increase the tolerance of the associated plants to the pollutant (Shim et al. 2000; Lodewyckx et al. 2001; Barac et al. 2004). In genetically engineered plant-associated bacteria, one or more genes may be inserted for enhanced remediation, for example, genes that code for biodegradative enzymes, biotic and abiotic stress, metal homeostasis, metal chelators and transporters, metal uptake regulators, and risk mitigation (Wu et al. 2006a,b; Doty 2008; Vangronsveld 2009; Weyens et al. 2013). Bacteria expressing a specific bacterial glutathione-S-transferase (GST), BphK^{LB400} [wild type and mutant (Ala180Pro)], isolated from *Burkholderia xenovorans* LB400 capable of dehalogenating toxic chlorinated organic pesticides, were shown to protect inoculated pea plants from the effects of a chlorinated organic pesticide, chlormequat chloride (McGuinness et al. 2007). Previously, it had been shown that mutating the conserved amino acid at position 180 in BphK^{LB400} from an alanine to proline residue resulted in an approximately twofold increase in GST activity toward a number of chlorinated organic substrates tested, including commonly used pesticides. These data suggest that BphK^{LB400} [wild type and mutant (Ala180Pro)], when inserted into endophytic or rhizospheric bacteria, could have potential for bioremediation of chlorinated organic pollutants in contaminatedsoil (McGuinness et al. 2006). Maintenance of the appropriate bacteria in contaminated areas can also be handicapped by the toxicity of the compounds. Transfer of catabolic plasmid NAH7 to *Pseudomonas putida* KT2440, an excellent rhizosphere colonizer, enables this strain to grow in naphthalene polluted soils without affecting the survival and colonization functions in corn (*Zea mays*) and cypress (*Cupressus sempervirens*) rhizospheres (Fernández et al. 2012). In a gene-expression experiment, it was demonstrated that the presence of naphthalene induced genes that encoded stress response and energy production proteins, regulators, transporters, and several other proteins of unknown function. Most of the stress-related genes were not induced in similar experiments carried out with the strain harboring the NAH7 plasmid. Construction of bacteria able to tolerate the toxicity of the contaminant is, therefore, another strategy to improve survival in the contaminated soil rhizosphere (Shim et al. 2000). Nevertheless, in the removal of contaminants from soils, the persistence of specific genes and not necessarily specific cells in the polluted environment is of utmost importance. Inoculation of poplar trees with the endophytic bacterium *Burkholderia cepacia* L.S.2.4 carrying the pTOM-Bu61 plasmid (that constitutively expresses the toluene degradation genes) improved the tree tolerance and evapotranspiration of toluene. Acquisition of the pTOM-Bu61 plasmid through horizontal gene transfer (HGT) by other endophytic bacteria better adapted to persist in the environment was the reason for the improvement (Weyens et al. 2011).

Sriprang et al. (2002) developed a phytobacterial system for the remediation of heavy metals (HMs) through the expression of tetrameric human metallothionein (MTL4) in *Mesorhizobium huakuii* sub sp. rengei B3, a strain which infects and forms nodules on a green manure. The resultant engineered bacteria, established symbiosis with *Astragalus sinicus*, produced metallothioneins (MTs), that is, metal binding proteins, which resulted in a 1.7–2.0-fold increase in Cd^{2+} accumulation. Ike et al. (2007) developed a novel bio-remediation system based on the symbiosis between leguminous plant and genetically engineered rhizobia. They desiged two types of recombinant rhizobia, carrying two genes, synthetic tetrameric metallothionein (AtMTL4) and cDNA encoding phytochelatin synthase (AtPCS) from *Arabidopsis thaliana*. Ike and his team incorporated AtMTL4 and AtPCS genes into *M. huakuii* sub sp. rengei B3. The resultant recombinant strains B3::nifHMTL4(pMPnifHPCS), and, B3(pMPnolBMTL4nifHPCS) accumulated 25- and 12-fold more Cd, respectively, in their cells. When these recombinant strains established a symbiotic relationship with *A. sinicus*, the accumulation of Cd in roots increased threefold. Further, the iron-regulated transporter 1 gene from *A. thaliana* (AtIRT1) was introduced into the previously mentioned AtMTL4 or ATPCS-recombinant rhizobium, *M. huakuii* subsp. strain B3 (Ike et al. 2007). When the recombinant strain established a symbiotic relationship with *A. sinicus,* there was a higher accumulation of Cu and As in the nodules compared with Cd and Zn. Yong et al. (2014) cloned phytochelatin synthase (PCS) genes from *Schizosaccharomyces pombe* that were expressed in *P. putida* KT2440. Phytochelatins (PCs) are cysteine-rich peptides with a high binding affinity for toxic heavy metal. Expressing the PC synthase gene (PCS) in plant growth-promoting bacteria may enhance its metal resistance and accumulation, consequently increasing phytoremediation efficiency in heavy metal pollution. The recombinant strain *P. putida* KT2440-SpPCS exhibited enhanced resistance to Hg, Cd, and Ag, and a three to fivefold increase in Cd accumulation along with an increase in nonprotein thiols, which led to an increase in the efficiency of phytoremediation of heavy metal contaminated environments. In addition, engineered bacteria resulted in significantly enhanced germination and growth of wheat, which indicates that *P. putida* KT2440 uses symbiosis for enhanced phytoremediation of heavy metals (Yong et al. 2014).

9.3 GENETICALLY ENGINEERED RHIZOBACTERIA IN PHYTOREMEDIATION OF ORGANIC AND INORGANIC ENVIRONMENTAL POLLUTANTS

Rhizospheric bacteria can be better equipped to colonize the plant rhizosphere and are the best option for degradation/detoxification of organic and inorganic pollutants and eco-restoration of contaminated environments (Glick 2003, 2012). Many endophytic bacteria help their host plant to overcome contaminant-induced stress responses, thus improving plant growth and development in contaminated environments (Van Aken et al. 2004a,b; van der Lelie et al. 2009; Weyens et al. 2010a,b). Research into how plant growth can be promoted has mainly concentrated on rhizobacteria, especially plant growth-promoting rhizobacteria. The first published report on the idea of engineering plant-associated bacteria for enhanced phytoremediation utilized rhizospheric rather than endophytic bacteria. Shim and colleagues engineered poplar rhizospheric bacteria for enhanced trichloroethylene (TCE) metabolism (Shim et al. 2000). Utilization of rhizobacteria isolated from the rhizosphere of poplar tree (*Populus canadensis* var. *eugene i* "Imperial Carolina"), already located in a contaminated area and re-inoculation after engineering those bacteria to carry toluene-o-monooxygenase (TOM) genes of *B. cepacia* strain G4, proved a better strategy for TCE removal than using rhizobacteria from other plants. Apart from good TCE degradation rates, the poplar-derived recombinants showed better rhizosphere competitiveness than other strains isolated from wheat (*Triticum aestivum*) or shrubs. Studies using two genetically modified strains of the rhizospheric bacteria *Pseudomonas fluorescens* F113, that is, *P. fluorescens* F113rifbph (with a single chromosomal insertion of the *bph* operon) (Brazil et al. 1995) and *P. fluorescens* F113: 1180 (with a single chromosomal insertion of the *bph* operon under the control of the *Sinorhizobium meliloti* node regulatory system) (Villacieros et al. 2005) reported that (i) the modified rhizospheric bacteria colonized roots as effectively as the wild type rhizospheric bacteria, (ii) *bph* genes were expressed *in situ* in soil, and (iii) the modified rhizospheric bacteria could degrade polychlorinated biphenyls(PCBs) more efficiently than the wild type rhizospheric bacteria, indicating considerable potential for the manipulation of the rhizosphere as a useful strategy for bioremediation. *P. fluorescens* F113: 1180 does not contain antibiotic resistance genes from the vector, making this strain more suitable for *in situ* applications. Since the *bph* element in *P. fluorescens* F113: 1180 is stable, lateral transfer of the *bph* element to a homologous recipient would not be expected to occur at detectable frequencies in the rhizosphere (Ramos et al. 1994). Wu et al. (2006a,b) demonstrated that the expression of a metal binding peptide (EC20) by recombinant rhizobacterium *P. putida* 06909 improved Cd binding and alleviated the cellular toxicity of Cd. Moreover, the inoculation of *Helianthus annuus* (sunflower) roots with engineered *P. putida* 06909 resulted in a significant decrease in Cd phytotoxicity and a 40% increase in Cd accumulation in plant roots. The symbiotic association distinctly improved the phytoextraction of heavy metals and enhanced the growth of plants, which suggests the use of genetically engineered plant growth-promoting bacteria in eco-restoration of contaminated sites.

9.4 GENETICALLY ENGINEERED ENDOPHYTIC BACTERIA IN PHYTOREMEDIATION OF ORGANIC AND INORGANIC ENVIRONMENTAL POLLUTANTS

Endophytic bacteria can be defined as bacteria that colonize the internal tissue of the plant without causing visible external signs of infection or a negative effect on the host. Endophytic bacteria are also known to have plant growth-promoting and pathogen control capabilities (Lodewyckx et al. 2001; Doty 2008; Weyens et al. 2009a,b). The majority of the world's known plant species are host to one or more endophytic species. During phytoremediation of organic contaminants, plants can further benefit from endophytes possessing appropriate degradation pathways and metabolic capabilities, leading to more efficient contaminant degradation and reduction of both phytotoxicity and evapotranspiration of volatile contaminants (Weyens et al. 2009c, 2013). Endophytic bacteria can be isolated from host plants of interest (e.g., plants native to a geographical region) and genetically enhanced to contain degradation pathways or genes to degrade target contaminants before being reinoculated back into the host plant for bioremediation purposes. Plant-associated endophytes genetically enhanced so as to degrade toxic organic compounds appear to offer more potential than rhizospheric bacteria for reducing phytotoxicity (Newman and Reynol 2005). Endophytes, by being inside of the plant, are protected from many of the problems encountered by rhizosphere bacteria in the soil, including competition from other soil bacteria; predation; and extremes of temperature, water content, oxygen, and pH. Nevertheless, endophytes utilize many of the same mechanisms to promote plant growth as rhizosphere bacteria (Weyens et al. 2009c). The most intensive study of a recombinant bacterial endophyte constructed for pest control purposes was performed with *Clavibacter xyli* subsp. cynodontis, expressing the *cryIA* gene from *Bacillus thuringiensis* (Lampel et al. 1994; Tomasino et al. 1995). In this study, the resultant protein was an endotoxin that is active against the European corn borer. The recombinant bacterium was effective in reducing insect boring in corn in laboratory, greenhouse, and field studies. However, significant increases in overall yield were not detected compared with plants that were not treated with the endophyte. Plant–endophyte partnerships can also be used for the cleanup of (ground) water contaminated with organic compounds. Pollutant-degrading genes can be incorporated in endophytic bacteria to improve plant–endophyte partnerships for the cleanup of water polluted with organic pollutants (Doty 2008; Weyens et al. 2009a,c,d). Barac et al. (2004) reported that a genetically enhanced endophytic strain of the soil bacterium *B. cepacia* G4 could increase inoculated yellow lupine plant tolerance to toluene and decrease phytovolatilization of toluene from the plant into the atmosphere by 50%–70% in laboratory-scale experiments. In this study, the plasmid pTOM, which encodes a pathway for the degradation of toluene, from a root-colonizing strain of the soil bacterium *B. cepacia* G4, was inserted into the yellow lupine (*Lupinus luteus* L.) endophyte *B. cepacia* BU0072 (derived from the endophytic strain B. cepacia L.S.2.4, and its toluene-degrading derivative VM1330) via bacterial conjugation, providing the genes for toluene degradation. Following the characterization of the resultant transconjugant strain, all three strains (i.e., the endophyte that could not degrade toluene, the root colonizer that could degrade toluene, a moderately hydrophobic [$logK_{ow}$ 2.69 at 20°C] volatile compound, and the transconjugant endophyte that could degrade toluene) were tested for the ability to degrade toluene in the presence of three-week-old yellow lupine plants. These tests showed that in the absence of added bacteria, increasing concentrations of toluene cause greater levels of phytotoxicity in plants. In contrast, when grown at toluene concentrations up to 1000 mg/L, plants that had been inoculated with the endophyte engineered to degrade toluene grew as well as the control plants that did not have toluene added. Moreover, the recombinant toluene-degrading endophytic strain not only protected its host against toluene phytotoxicity but also lowered the amount of evapotranspiration of toluene from the plant into the atmosphere by 50%–70% in laboratory-scale experiments. A major concern of phytoremediation systems working on volatile solvent contamination has been the fact that the plants can transpire these compounds through the leaf stomata or stem lenticels. Thus, this decrease in transpired toluene could prove to be significant. Later, Taghavi et al. (2009) extended this work to poplar tree (*Populus trichocarpa deltoides* cv. Hoogvorst) and showed that this degradative plasmid, pTOM-Bu61, could transfer naturally, via HGT, to a number of different endophytes *in planta*, promoting more efficient degradation of toluene in poplar plants. HGT results in the natural endophyte population having the capacity to degrade environmental pollutants without the need to establish the inoculant strain long term. Endophytes that have been engineered by HGT have the distinct advantage that they may not be considered genetically modified microorganisms (GMMs) and could, therefore, be exempt from current international and national GM legislation, thus facilitating the testing of these microorganisms in the field at an accelerated pace. The first *in situ* inoculation of poplar trees growing on a TCE-contaminated site with TCE-degrading strain *Pseudomonas putida* W619-TCE was done by Weyens et al. (2010b). Bacterial application enhanced plant growth and reduced the accumulation of TCE in the shoots and leaves of poplar plants. The inoculation resulted in a 90% reduction of TCE evapotranspiration under the field conditions. The inoculation of a genetically modified endophytic bacterium was also proven to be useful to enhance degradation of aromatic and volatile organic compounds present in the water. These studies proposed that engineered endophytic bacteria can also be inoculated to plants to enhance remediation of water polluted with organic compounds. Another study demonstrated the use of engineered endophytes for improving phytoremediation of environments contaminated by organic pollutants and toxic metals (Weyens et al. 2011). Yellow lupine was inoculated with *B. cepacia* VM1468 possessing (i) the pTOM-Bu61 plasmid coding for constitutive TCE degradation and (ii) the *ncc-nre* Ni resistance/sequestration. Inoculation with *B. cepacia* VM1468 into plants resulted in a decrease in Ni and TCE phytotoxicity, which was reflected by a 30% increase in root biomass and up to a 50% decrease in the activities of enzymes involved in antioxidative defense in the roots. Germaine et al. (2006) described inoculation of pea plants (*Pisum sativum*) with *P. putida* VM1450, a genetically modified bacterial endophyte that naturally possesses the ability to degrade a organochlorine herbicide, 2,4-dichlorophenoxyacetic acid (2,4 D). The inoculated plants had a higher degradation capacity of up to 40% for 2,4 D from the soil. In a another separate study, the same authors showed that inoculation of pea plants with an endophytic naphthalene-degrading bacterium, *P. putida* VM1441, protected the host plant from phytotoxic effects of naphthalene, increased seed germination, increased transpiration rates, and removed 40% more naphthalene from artificially contaminated soils than did uninoculated plants (Germaine et al. 2009. The work of Barac et al. (2004) raises the possibility of introducing a variety of potentially useful genes into endophytes. For example, most plants respond to a wide range of different stresses, including the presence of phytopathogens, high or low temperature, drought, flooding, high salt, metals, xenobiotic organic compounds, and wounding by synthesizing "stress ethylene." The increased amounts of ethylene formed in response to trauma can be both the cause of some of the symptoms of stress and the inducer of stress responses that enhance survival of the plant under adverse conditions. It has been shown that the enzyme 1-aminocyclopropane-1-carboxylic acid (ACC) deaminase, when present in plant growth-promoting bacteria, including endophytes, can act to lower the level of ethylene in a plant (Glick 2012, 2014). Treatment of plants with bacteria that contain this enzyme can also lead to the synthesis of much less ethylene than normal in response to environmental stress, thereby decreasing the stress-induced damage to the plant. Thus, to engineer bacteria that not only facilitate the degradation of a target organic compound, such as toluene, but also decrease the stress that a plant experiences in the presence of a xenobiotic, it may be necessary to ensure that the endophyte that is used can produce the enzyme ACC deaminase (Vangronsveld 2009; Weyens et al. 2009a,b,c,d). Given the current political impediments around the world to using either genetically modified plants or genetically modified bacteria in the environment, endophytes that have been engineered by plasmid transfer, and possibly also selected for specific enzymatic activities, such as ACC deaminase, would not be considered genetically modified organisms (GMOs) in many countries. Therefore, the type of engineered strain described by Barac et al. (2004) might be exempt from much of the current legislation. This, along with the technical efficacy of this approach, could help this technology be rapidly adopted and put into widespread use.

The concept of engineering endophytes rather than rhizospheric bacteria for phytoremediation was also tested for enhancing remediation of metals. Genes for Ni tolerance (*ncc* [nickel-cadmium-cobalt resistance]-*nre*) were transferred from *Ralstonia metallidurans* 31A into two endophytic strains, *B. cepacia* L.S.2.4 and *Herbaspirillum seropedicae* LMG2284 (Lodewyckx et al. 2001). As with the TOM system, the Ni tolerance genes needed to be integrated into the chromosome of the endophytes using a transposon, because plasmids were unstable. *B. cepacia* and *H. seropedicae* containing the *ncc-nre* Ni resistance system were inoculated onto seeds of their host plants, *Lupinus luteus* and *Lolium perenne*. Contrary to expectation, when the plants were grown in a medium containing 0.25 mM $NiCl_2$, the biomass, root length, and shoot length were all reduced whether or not the endophyte contained the nickel resistance genes. Although the modified endophytes themselves had increased resistance to the toxic effects of Ni, they apparently did not influence the growth of the plants or cause an increased translocation of Ni by the inoculated plants. In a follow-up to this work, Weyens et al. (2009c) showed that endophytic bacteria carrying an appropriate degradation pathway can be used successfully in the field together with poplar trees to degrade TCE. Finally, when an endophyte is engineered by HGT of a plasmid containing a specific degradative pathway, the deliberate release of the engineered bacterial strain is not restricted under European regulations (Weyens et al. 2009a,d). Wang et al. (2010) successfully colonized *B. cepacia* strain FX2, which contains plasmids with the gene encoding for catechol 2,3-dioxygenase (a key enzyme in the degradation pathway of monocyclic aromatic compounds), in corn and wheat to promote crop growth and reduce evapotranspiration.

9.5   CHALLENGES AND CONCERNS OF USING GENETICALLY ENGINEERED BACTERIA IN ASSISTING PHYTOREMEDIATION

Engineering plant-associated microbes is of limited scope, as these mainly concentrate around the roots, show limited distribution outside the rhizosphere, and depend chiefly on the host plants (Wu et al. 2006a,b). Hence, their controlled use outside the rhizosphere is not an easy task. Not every bacterium with the necessary pollutant-degrading capacity has the ability to grow well within the plant species where the contamination is present (Ojuederie and Babalola 2017). Horizontal gene transfer of environmental plasmids is a natural process, and its utilization in rhizoremediation strategies could be a high-quality method to overcome the legal issues concerning gene transfer. Still, release of GMOs for the removal of contaminants, although an excellent biotechnological approach, is limited by legal and ethical constraints in many countries, which is a drawback for the use of transgenic organisms, and this, together with some well-sustained scientific concerns, may limit the development of this field.

10 Success Stories of Microbe-Assisted Phytoremediation

10.1 INTRODUCTION

Soil, air, and water pollution have become a global problem due to unprocessed emission of toxic heavy metals into the environment (Roane and Pepper 2000; Yadav et al. 2014; Suman et al. 2018). Over the past few decades, there has been avid interest in developing *in situ* strategies for remediation of environmental contaminants, especially persistent organic chemicals and heavy metals. Phytoremediation is a relatively new technology that offers clear advantages over traditional methods for site cleanup (Chaney 1983; Cunningham 1995, 1997; Marques et al. 2009; Cabral et al. 2015). This plant based technology has been used to treat many classes of contaminants, including petroleum hydrocarbons, chlorinated solvents, pesticides, explosives, heavy metals, and radionuclides, in soil and polluted water or wastewater (Nedelkoska and Doran 2000; Schäffner et al. 2002; Cang et al. 2011; Meier et al. 2012; Luo et al. 2014; Bhargavi et al. 2015). Microbe-assisted phytoremediation, a specific strategy of phytoremediation that involves both plants and their associated rhizosphere microbes, can occur naturally or can be actuated by deliberately introducing specific microbes (Weyens et al. 2009a,b; Rajkumar et al. 2012; Nanekar et al. 2015; Rajtor et al. 2016; Thijs et al. 2017). These microbes can be contaminant degraders and/or can promote plant growth under stress conditions (Jing et al. 2007; Phillips et al. 2008; Khan et al. 2014a,b; Yang et al. 2016; Ahemad 2019). Because initial phytoremediation research showed great promise as a cost-effective remedial strategy, considerable effort has been devoted to making the transition from the laboratory to commercialization. Some of its applications have only been assayed at the laboratory or greenhouse level, but others have been field tested sufficiently to allow full-scale operation (Teng et al. 2010; Huo et al. 2012; Khan et al. 2014a,b; Ijaz et al. 2015; Ashraf et al. 2018a,b; Hussain et al. 2018). Some case studies for microbe-assisted phytoremediation in India and abroad are briefly described in the present chapter.

10.2 PHYTOREMEDIATION IN LAB-SCALE/MICROCOSM STUDY

Prior to phytoremediation field trials, extensive research was performed in laboratories and greenhouses. Some of this work explored the effects of plants on removal of contaminants from spiked soil and soil excavated from contaminated sites. Many of these experiments provided valuable insights into the types and specific mechanisms of phytoremediation of organic contaminants.

Case Study 1: Understanding Plant–Microbe Interactions for Phytoremediation of Petroleum-Polluted Soil (Nie et al. 2011)

Plant–microbe interactions are considered important processes determining the efficiency of phytoremediation of petroleum pollution. Nie et al. (2011) performed an experiment using a microcosm approach to examine how plant eco-physiological traits, soil nutrients, and microbial activities were influenced by petroleum pollution in *Phragmites australis*. Generally, petroleum pollution reduced plant performance, especially at early stages of plant growth, and showed negative effects on the net accumulation of inorganic nitrogen from its organic forms. In order to overcome initial deficiency of inorganic nitrogen, plants by dint of high colonization of arbuscular mycorrhizal fungi (AMF) might absorb some dissolved organic nitrogen for their growth in petroleum-polluted soils. In addition, through using a real-time polymerase chain reaction (RT-PCR) method, authors quantified hydrocarbon-degrading bacterial traits based on their catabolic genes (i.e., *alkB* (alkane monooxygenase), *nah* (naphthalene dioxygenase), and *tol* (xylene monooxygenase genes). This enumeration of target genes suggests that different hydrocarbon-degrading bacteria experienced different dynamic changes during phytoremediation and a greater abundance of *alkB* was detected during vegetative growth stages. The information provided by this study enhances our understanding of the effects of petroleum pollution on plant–microbe interactions and the roles of these interactions in the phytoremediation of petroleum-polluted soil.

Case Study 2: Microbe-Assisted Phytoremediation of Oil Sludge and Role of Amendments: A Mesocosm Study (Nanekar et al. 2015)

A mesocosm study was conducted by Nanekar and his colleagues to elucidate the influence of amendments such as microbial consortium, plants (*Vetiveria zizanioides*), bulking agents (wheat husk), and nutrients on remediation of oil sludge. During the experiment, oil degradation was estimated gravimetrically and polyaromatic hydrocarbons (PAHs) were quantified on gas chromatography–mass spectrometry (GC–MS). Additionally, dehydrogenase activity was also monitored. The treatment integrated with bulking agents, nutrients, consortium, and plants resulted in increased dehydrogenase activity and complete mineralization of higher PAHs. Furthermore, 72.8% total petroleum hydrocarbon (TPH) degradation was observed in bulked treatment with plants, nutrients, and consortium followed by 69.6% and 65.4% in bioaugmented treatments with and without nutrients, respectively, as compared to control. This study validates a holistic approach for remediation of oil sludge–contaminated sites by the use of microbe-assisted phytoremediation technology, which not only solves the problem of oil contamination but also takes care of heavy metal contamination.

Case Study 3: Bacterial-Assisted Phytoremediation for Enhanced Degradation of Highly Sulfonated Diazo Reactive Dye (Khandare et al. 2012)

The purpose of this work was to explore plant and bacterial synergism for enhanced degradation of Remazol Black B (RBB) dye, which are toxic, carcinogenic, and mutagenic from the effluent. The developed bacterial consortium ZE was found to be more efficient than individual plants and bacteria. Also, *Zinnia angustifolia* roots showed significant induction in lignin peroxidase, laccase, DCIP (2,6-dichlorophenol-indophenol) reductase, and tyrosinase activities during dye decolorization. *Exiguobacterium aestuarii* strain ZaK showed significant induction in the activities of veratryl alcohol oxidase, azo reductase, and DCIP reductase. The phytotoxicity study revealed the nontoxic nature of the metabolites formed after dye degradation. Authors concluded that consortium ZE was found to be more efficient and faster in the degradation of RBB when compared to degradation by *Z. angustifolia* and *E. aestuarii* individually.

Case Study 4: Rhizobacteria (*Pseudomonas* sp. SB)-Assisted Phytoremediation of Oily Sludge–Contaminated Soil by Tall Fescue (*Testuca arundinacea* L.) (Liu et al. 2013)

The biosurfactant-producing *Pseudomonas* sp. SB isolated from rhizosphere of *T. arundinacea* L. grown in a petroleum-contaminated soil has plant growth-promoting traits and found a suitable bioinoculant to assist phytoremediation of oily sludge–contaminated soils (Liu et al. 2013). Besides biosurfactant-producing properties, *Pseudomonas* sp. SB also produced indole-3-acetic acid (IAA), siderophores, and 1-aminocyclopropane-1-carboxylic acid (ACC) deaminase. The inoculation of *Pseudomonas* sp. SB promoted the growth of *T. arundinacea* L. and significantly enhanced the degradation of TPH and PAHs. It also increased the microbial activity and diversity in the soil. Furthermore, the inoculation of *Pseudomonas* sp. SB markedly decreased the toxicity of the contaminated soil.

Case Study 5: Plant Growth-Promoting Rhizobacteria (PGPR)-Enhanced Phytoremediation of Petroleum-Contaminated Soil and Rhizosphere Microbial Community Response (Hou et al. 2015)

An experiment was conducted by Hou and his team in 2015 to investigate petroleum phytoremediation enhancement by two PGPR strains, specifically the correlation between petroleum hydrocarbon (PH) fractions and bacterial community structure affected by remediation and PGPR inocula. PGPR inoculation increased *T. arundinacea* L. biomass and PH was removed in all treatments. Maximum PH removal, particularly high-molecular-weight aliphatic hydrocarbons and PAHs, was observed in *T. arundinacea* L. inoculated with PGPR. Further, changes in bacterial community structure analyzed by high-throughput pyrosequencing of 16 s rRNA gene showed that the relative abundance of phyla Gammaproteobacteria and Bacteroidetes were increased after different treatments compared with controls. Moreover, a bacterial guild mainly comprising the genera *Lysobacter, Pseudoxanthomonas, Planctomyces, Nocardioides, Hydrogenophaga,* and *Ohtaekwangia* was found to be positively correlated with aliphatic PH fraction removal by redundancy analysis (RDA) analysis, implying that PH degradation was unrelated to bacterial community diversity but positively correlated with specific petroleum degraders and biosurfactant producers.

Case Study 6: Characterization of Heavy Metal–Resistant Endophytic Bacteria from Rape (*Brassica napus*) Roots and Their Potential in Promoting the Growth and Lead (Pb) Accumulation of Rape (Sheng et al. 2008a)

Sheng et al. (2008a) conducted a pot experiment for investigating the capability of the two Pb-resistant endophytic bacteria, *Pseudomonas fluorescens* G10 and *Microbacterium* sp. G16, isolated from *B. napus* roots grown in HM-contaminated soils to promote the growth and Pb uptake of rape from Pb-amended soil (Sheng et al. 2008a). These two strains could colonize the root interior and rhizosphere soil of *B. napus* after root inoculation. Endophytic bacteria strains G10 and G16 exhibited various multiple heavy metal and antibiotic resistance characteristics and increased water-soluble Pb in solution and in Pb-added soil. These two endophytic bacteria *strains* colonized in the plant tissue interiors protect plants against the inhibitory effects of high concentrations of Pb and promote the plant growth by production of IAA or ACC deaminase, and the endophytic bacteria colonizing in the rhizosphere soils of rape might promote plant growth and Pb uptake by production of IAA, siderophore or ACC deaminase or by solubilization of Pb in soils as shown in our research.

Case Study 7: Effect of PGPR and Arbuscular Mycorrhizal Fungi (AMF) Inoculation on Oats in Saline-Alkali Soil Contaminated by Petroleum to Enhance Phytoremediation (Xun et al. 2015)

To investigate the effect of PGPR and AMF on phytoremediation in saline-alkali soil contaminated by petroleum, a pot experiment with oat plants (*Avena sativa*) was conducted under greenhouse conditions for 60 days (Xun et al. 2015). The result demonstrated that petroleum inhibited the growth of the plants; however, inoculation with PGPR in combination with AMF resulted in an increase in dry weight and stem height compared with non-inoculated controls. Petroleum stress increased the accumulation of malondialdehyde (MDA) and free proline and the activities of antioxidant enzymes such as superoxide dismutase, catalase, and peroxidase. Application of PGPR and AMF augmented the activities of three enzymes compared to their respective uninoculated controls but decreased the MDA and free proline contents, indicating that PGPR and AMF could make the plants more tolerant to harmful hydrocarbon contaminants. It also improved the soil quality by increasing the activities of soil enzymes such as urease, sucrase, and dehydrogenase. In addition, the degradation rate of TPH during treatment with PGPR and AMF in moderately contaminated soil reached a maximum of 49.73%. Therefore, the authors concluded plants treated with a combination of PGPR and AMF had a high potential to contribute to remediation of saline-alkali soil contaminated with petroleum.

Case Study 8: A Comparative Study to Evaluate Natural Attenuation, Mycoaugmentation, Phytoremediation, and Microbial-Assisted Phytoremediation Strategies for the Bioremediation of an Aged PAH-Polluted Soil (García-Sánchez et al. 2018)

A pot experiment was conducted to comparatively evaluate four different strategies, including natural attenuation (NA) and mycoaugmentation (M), by using *Crucibulum laeve*, phosphorous (P)-using maize plants, and microbe-assisted phytoremediation for the bioremediation of an aged PAH-polluted soil at 180 days (García-Sánchez et al. 2018). The P treatment had higher affinity degrading 2–3 and 4-ring compounds than NA and M treatments, respectively. However, M and P treatments were more efficient with regard to naphthalene, indeno [*l*,2,3-*c*,*d*] pyrene, and benzo [*g*,*h*,*i*] perylene degradation with respect to NA. However, 4, 5–6 rings underwent a strong decline during microbe-assisted phytoremediation, which was the treatment that showed the highest rates of PAH degradation. Sixteen PAH compounds, except fluorene and dibenzo [*a*,*h*] anthracene, were found in maize roots, whereas naphthalene, phenanthrene, anthracene, fluoranthene, and pyrene were accumulated in the shoots in both P and MAP treatments. However, higher PAH content in maize biomass was achieved during MAP treatment with respect to P treatment. The bioconversion and translocation factors were <1, indicating that phytostabilization/phytodegradation processes occurred rather than phytoextraction. The microbial biomass, activity, and ergosterol content were significantly boosted in MAP treatment with respect to the other treatments. The authors demonstrated that maize-*C. laeve* association was the most profitable technique for the treatment of an aged PAH-polluted soil when compared to other bioremediation approaches.

Case Study 9: AMF–Assisted Phytoremediation of Pb-Contaminated Site (Schneider et al. 2016)

Knowledge of the behavior of plant species associated with AMF and the ability of such plants to grow on metal-contaminated soils is important to phytoremediation. Schneider et al. (2016) evaluated the occurrence and diversity of AMF and plant species as well as their interactions in soil contaminated with Pb. Thirty-nine AMF species from 6 families and 10 genera was identified. The *Acaulospora* and *Glomus* genera exhibited the highest occurrences both in bulk and rhizosphere soils. All of the herbaceous species presented mycorrhizal colonization. The highest Pb concentrations (mg kg^{-1}) in roots and shoots, respectively, were observed in *Vetiveria zizanioides* (15 433 and 934), *Pteris vittata* (9343 and 865), *Pteridium aquilinum* (1433 and 733), and *Ricinus communis* (1106 and 625). Such knowledge can aid in developing soil phytoremediation techniques such as phytostabilization.

Case Study 10: *Streptomyces pactum*–Assisted Phytoremediation in Zn/Pb Smelter-Contaminated Soil and Its Impact on Enzymatic Activities (Ali et al. 2017)

To assist phytoremediation by sorghum in soil contaminated by smelters/mines in Feng County (FC), a pot experiment was performed to examine the phytoremediation potential of *S. pactum* (Act12)+biochar. The results showed that root uptake of Zn and Cd were reduced by 45% and 22%, respectively, whereas the uptake of Pb and Cu increased by 17% and 47%, respectively. The shoot and root dry weight and chlorophyll content improved after Act12 inoculation. β-glucosidase, alkaline phosphatase, and urease activities in soil improved and antioxidant activities decreased after application of Act12+biochar due to a reduction in stress from potentially toxic trace elements (PTEs). Bio-concentration factor, translocation factor, and metal extraction amount confirmed the role of Act12 in the amelioration and translocation of PTEs. Overall, Act12 promoted the phytoremediation of PTEs (Ali et al. 2017).

Case Study 11: Fungi-Assisted Phytoextraction of Pb: Tolerance, Plant Growth-Promoting Activities, and Phytoavailability (Manzoor et al. 2019)

A pot experiment was done to explore the potential of different Pb-resistant fungal strains to promote phytoextraction of Pb-contaminated soils (Manzoor et al. 2019). In this experiment, five nonpathogenic fungal strains (*Trichoderma harzianum, Penicillium simplicissimum, Aspergillus flavus, Aspergillus niger,* and *Mucor* spp.) were tested to check their ability to modify soil properties and to increase Pb phytoavailability at varying concentrations. Pb tolerance of fungal strains followed a decreasing order as *A. niger* > *T. harzianum* > *A. flavus* > *Mucor* sp. > *P. simplicissimum.* However, Pb solubility induced by *A. flavus* and *Mucor* spp. was increased by 1.6- and 1.8-fold, respectively, as compared to the control soil (Pb added, without fungi). *A. flavus* and *Mucor* spp. lowered the soil pH by 0.14 and 0.13 units, in soils spiked with 2000 mg Pb kg^{-1}. The maximum increase in the percentage of organic matter (OM) recorded was for *A. flavus* at 500 mg Pb kg^{-1} soil. Plant growth-promoting assays confirmed the beneficial role of these fungal strains. Significantly high production of IAA and siderophores was observed with *A. niger* and phosphate solubilization with *P. simplicissimum.* Based on the results in Pb-contaminated soils, *Pelargonium hortorum* L. inoculated with *Mucor* spp. showed the potential to enhance phytoextraction of Pb by promoting Pb phytoavailability in soil and improving plant biomass production through plant growth-promoting activities.

10.3 PHYTOREMEDIATION IN THE FIELD

Studies with plant species potentially suitable for microbe-assisted phytoremediation are widely represented in scientific literature. However, the in-depth understanding of the biological processes associated with the reintroduction of indigenous bacteria and plants and their performance in the degradation of organic pollutants is still the limiting step for the application of these bioremediation solutions in a field context. Although the successes of laboratory and greenhouse experiments are often difficult to replicate in the field, a number of encouraging results from the field have been reported. Despite our understanding of the mechanisms of remediation, and the success of studies in the laboratory and greenhouse, efforts to translate phytoremediation research to the field have proven challenging. Although there have been many encouraging results in the past decade, there have also been numerous inconclusive and unsuccessful attempts at phytoremediation in the field.

Case Study 1: Influence of Arbuscular Mycorrhiza and *Rhizobium* on Phytoremediation by Alfalfa of an Agricultural Soil Contaminated with Polychlorinated Bipehnyl (PCB): A Field Study (Teng et al. 2010)

A field experiment was conducted by Teng et al. (2010) to study the effects of inoculation with the AMF *Glomus caledonium* and/or *Rhizobium meliloti* on phytoremediation of an agricultural soil contaminated with weathered PCB by alfalfa grown for 180 days. Planting alfalfa (P), alfalfa inoculated with *G. caledonium* (P+AM), alfalfa inoculated with *R. meliloti* (P+R), and alfalfa co-inoculated with *R. meliloti* and *G. caledonium* (P+AM+R) significantly decreased initial soil PCB concentrations by 8.1, 12.0, 33.8, and 43.5%, respectively. Inoculation with *R. meliloti* and/or *G. caledonium* (P+AM+R) increased the yield of alfalfa and the accumulation of PCBs in the shoots. Soil microbial counts and the carbon utilization ability of the soil microbial community increased when alfalfa was inoculated with *R. meliloti* and/or *G. caledonium*. Results of this field study suggest that synergistic interactions between AMF and *Rhizobium* may have great potential to enhance phytoremediation by alfalfa of an agricultural soil contaminated with weathered PCB (Teng et al. 2010).

Case Study 2: Large-Scale Remediation of Oil-Contaminated Water Using Floating Treatment Wetlands (Afzal et al. 2019)

Floating treatment wetlands (FTWs) are a sustainable approach for remediating contaminated water. In a large-scale remediation study, authors used four different plants, *Phragmites australis*, *Typha domingensis*, *Leptochloa fusca*, and *Brachiaria mutica*, to vegetate a floating mat with an area of 3058 m^2 (Afzal et al. 2019). The FTWs constructed in this manner were used to treat an oil-contaminated water stabilization pit. The plants and the water in the pit were inoculated with a consortium of 10 different hydrocarbon-degrading bacteria (*Ochrobactrum intermedium* R2, *Microbacterium oryzae* R4, *Pseudomonas aeruginosa* R25, *P. aeruginosa* R21 *Acinetobacter* sp. LCRH81, *Klebsiella* sp. LCRI-87, *Acinetobacter* sp. BRSI56, *P. aeruginosa* BRRI54 (isolated from the shoot and root interior of *Brachiaria mutica*, respectively), *Bacillus subtilus* LORI66 and *Acinetobacter junii* TYRH47. These strains are able to degrade hydrocarbons, produce biosurfactants, and promote plant growth. The application of FTWs to the pit reduced chemical oxygen demand (COD), biochemical oxygen demand (BOD), total dissolved solids (TDS), hydrocarbon content, and heavy metals by 97.4%, 98.9%, 82.4%, 99.1%, and 80%, respectively, within 18 months. All plants survived and showed healthy growth, but maximum development and biomass production were exhibited by *P. australis*. Moreover, the bacteria used for inoculation were able to persist and show degradation activity in the water as well as in the rhizoplane, roots, and shoots of the plants. Afzal et al. (2019) concluded that FTWs can be applied to oil-contaminated water stabilization pits for affordable and effective water treatment.

Case Study 3: Phytoremediation Assisted by Mycorrhizal Fungi of a Mexican Defunct Lead-Acid Battery Recycling Site (González-Chávez et al. 2019)

A field experiment was conducted over 15 months to study the effects of four AMF on the growth of *Ricinus communis* accession SF7. Plants were established on amended soil (vermicompost:sawdust:soil 1:1:1) severely polluted by lead-acid batteries (LABs). Plants inoculated with *Acaulospora* sp., *Funneliformis mosseae*, and *Gigaspora gigantea* had 100% survival in comparison to noninoculated plants (57%). These same AMF enhanced palmitic and linoleic acid content in seeds of *R. communis*. *Acaulospora* sp. modified rhizosphere soil pH and decreased Pb foliar concentrations 3.5-fold, while *F. mosseae* BEG25 decreased Pb soil availability three times in comparison to noninoculated plants. Spatial changes in Pb soil availability were observed at the end of this research. No fungal effect on P, Ca, Cu foliar concentrations, soluble sugars, proline, chlorophyll, or the activity of two oxidative stress enzymes was observed. Mycorrhizal colonization from the inoculated fungi was between 40% and 60%, while colonization by native fungi was between 16% and 22%. A similar percentage of foliar total phenolic compounds were observed in non-mycorrhizal plants and those inoculated with *G. gigantea* and *Acaulospora* sp. This research reports effects of AMF on *R. communis* (castor bean) shrubs when grown on a LAB recycling site, suggesting the use of *Acaulospora* sp. and *F. mosseae* BEG25 in phytostabilization to ameliorate Pb pollution and decrease its ecological risk.

Case Study 4: Hydrocarbon Degradation Potential and Activity of Endophytic Bacteria Associated with Prairie Plants (Phillips et al. 2008)

Phillips et al. (2008) evaluated the dominant bacterial endophytes in five plant species at long-term phytoremediation field sites. *Pseudomonas* spp.-dominated endophytic communities experienced increased alkane hydrocarbon degradation potential and activity, whereas *Brevundimonas*- and *P. rhodesiae*-dominated endophytic communities were associated with increased PAH degradation potential and activity. Then the authors suggested that heterogeneously distributed endophytic microbial populations might impact the ability of plants to promote the degradation of specific types of hydrocarbons. The authors' results show that diverse plant species growing in weathered hydrocarbon–contaminated soil maintain distinct, heterogeneously distributed endophytic microbial populations, which may impact the ability of plants to promote the degradation of specific types of hydrocarbons.

Case Study 5: *In Situ* Phytoremediation of PAH-Contaminated Soil by Intercropping Alfalfa (*Medicago sativa* L.) with Tall Fescue (*Festuca arundinacea* Schreb.) and Associated Soil Microbial Activity (Sun et al. 2011)

A 7-month field experiment was conducted to investigate the PAH remediation potential of two plant species (*M sativa* L. and *Festuca arundinacea* Schreb.) and changes in counts of soil PAH-degrading bacteria and microbial activity (Sun et al. 2011). Results revealed that the average removal percentage of total PAHs in intercropping (30.5%) was significantly higher than in monoculture (19.9%) or unplanted soil (−0.6%). About 7.5% of 3-ring, 12.3% of 4-ring, and 17.2% of 5(+6)-ring PAHs were removed from the soil by alfalfa, with corresponding values of 25.1%, 10.4%, and 30.1% for tall fescue. Intercropping significantly enhanced remediation efficiency. About 18.9% of 3-ring, 30.9% of 4-ring, and 33.4% of 5(+6)-ring PAHs were removed by the intercropping system. Intercropping systems can increase soil PAH-degrading bacterial counts and microbial activities, suggesting that alfalfa and tall fescue growing together can restore the microbiological functioning of PAH-contaminated soil. The authors concluded cropping promoted the dissipation of soil PAHs. Tall fescue gave greater removal of soil PAHs than alfalfa, and intercropping was more effective than monoculture.

Although microbe-assisted phytoremediation is relatively straightforward to implement, there are numerous environmental factors that can positively or negatively influence this remedial strategy in the field. These include daylight hours, length of growing season, weather, soil structure and composition (organic matter and nutrient content, grain size distribution), natural soil chemistry (high salt, low or high pH), availability of water and oxygen in the soil, and effects of weathering on organic pollutants. Each site is unique, and the various influences result in complex and dynamic scenarios, particularly within the rhizosphere. The key to success is to be aware of the variables, their potential effects on the phytoremediation outcome, and the steps that can be taken to mitigate negative influences and maximize positive ones. Although the successes of laboratory and greenhouse experiments are rarely realized to the same extent in the field, there have been encouraging results that justify the continued use of phytoremediation (Rojjanateeranaj et al. 2017).

11 Emerging Issues and Challenges of Microbe-Assisted Phytoremediation

11.1 INTRODUCTION

Microbe-assisted phytoremediation is a technology that is based on the combined action of plants and their associated microbial communities to degrade, remove, transform, or immobilize toxic compounds located in soils, sediments, groundwater, and surface water (Phillips et al. 2008; Germaine et al. 2009; Ahemad 2015; González-Chávez et al. 2019; Manzoor et al. 2019). This plant microbe–based technology has been used to treat many classes of environmental contaminants, including melanoidins, petroleum hydrocarbons, chlorinated solvents, pesticides, explosives, heavy metals, and radionuclides (Weyens et al. 2009a,b,c; Liu et al. 2013; Nanekar et al. 2015; Schneider et al. 2016; Ali et al. 2017; García-Sánchez et al. 2018; Hrynkiewicz et al. 2018). Candidate plants for phytoremediation should have characteristics such as high biomass production, extensive root systems, and ability to tolerate high concentrations of pollutants and withstand environmental stress (Baker and Brooks 1989; Cunningham and Berti 1993; Cunningham et al. 1995; Garbiscu and Alkorta 2001; Karthikeyan and Kulakow 2003). Microbe-assisted phytoremediation is a technology whose time has nearly arrived. In the past ten years, scientists have developed a much better understanding of precisely how various microbes contribute to phytoremediation, and the efficacy of these approaches has been demonstrated under laboratory and field conditions as well (Glick 2003; Ma et al. 2016a,b; Kong and Glick 2017). In the present chapter, we discuss the emerging issues and challenges of plant microbe–based remedial strategies for *in situ* removal of organic and inorganic pollutants from contaminated sites. Further, the future prospects of plant microbe–based phytoremediation technology are highlighted.

11.2 ADVANTAGES AND CHALLENGES/LIMITATION OF MICROBE-ASSISTED PHYTOREMEDIATION TECHNOLOGY

There are numerous advantages of microbe-assisted phytoremediation technology that other remedial options do not provide, which can be summarized as follows:

- It is cost effective in comparison with the traditional remediation technologies, eco-friendly, has low disruptiveness to the environment and a landscape-friendly nature, and the potential to remediate a wide range of environmental pollutants.
- Plant growth results in net carbon sequestration (greenhouse gas storage): approximately 6 tons/hectares annually, which adds more environmental appeal to this already "green" technology.
- Green technologies have high public acceptance, making microbe-assisted phytoremediation an attractive option for environmentalists, industry, policy makers, and regulators.
- As a typical *in situ* technique, it may be applied to a broad range of contaminated sites, even those previously neglected due to the high cost of conventional remediation technologies.
- Plants, as autotrophic systems with large biomass, require only modest nutrient input, and their roots help to stabilize soil, which prevents the spread of contaminants through water and wind erosion.
- Microbe-assisted phytoremediation can be used in any geographical area that can support plant growth, including much of the subarctic.
- It can be used effectively at remote sites where alternative means of persistent organic pollutant remediation would be cost prohibitive.
- Another benefit of this plant-microbe technology is that the overall soil quality and structure improve at remediated sites due to organic materials, nutrients, and oxygen that are added to soil via plant and microbial metabolic processes.
- Plants can be easily grown without much effort, and growth of plants is driven by solar energy. Plant growth and maintenance costs after initial site preparation and planting are minimal.

Despite the numerous advantages to using microbe-assisted phytoremediation, there are also the following challenges that limit the use of this technology for cleanup of environmental pollutants.

- A frequently cited challenge of microbe-assisted phytoremediation technology, compared to strategies such as excavation and *ex situ* treatment, is that the rate of remediation is much slower. It may take several years for soil and wastewater remediation.
- It is an applicable approach for eco-restoration of sites that have low to moderate levels of metal or organic/metal-organic pollution due to unsustainable plant-microbe growth in highly contaminated soils. Therefore, microbe-assisted phytoremediation can be effective in sites with relatively low pollution levels and is restricted to places where contaminants are reachable for the plant roots.
- Phytoremediating plants are often unique ecotypes that inhabit specific ecosystems, and their cultivation in other environmental conditions is difficult. Plant-microbe remediation potential is also limited in unfavorable environmental conditions.

- A significant number of plants have been tested for their ability to accumulate vast quantities of heavy metals; however, phytoextraction efficacy of most metal hyperaccumulator plants is generally restricted by their low biomass and slow growth rate, and they do not produce sufficient biomass to make this process efficient in the field.
- Plants with low biomass yields and reduced root systems do not prevent the leaching of contaminants into aquatic systems.
- The metal accumulation capacity of some plants may be compromised and ineffective due to pest and disease attack in climate-affected tropical and subtropical regions.
- Improper biomass disposal of plants after phytoextraction may increase the risk of food chain contamination with heavy metals.
- The growth of plants during phytoremediation leads to changes in physic-chemical properties of soil that can make metals more bioavailable to the food chain before they can be remediated, thus posing environmental risks that can negate some of the positive effects of this technology.
- Metals that are phytostabilized are not removed from the soil, and ever-changing soil conditions may lead to contaminant re-release in the environment; hence, long-term monitoring of the site may be required to avoid land use changes in the future.
- The inability of introduced microbes to compete with existing microflora and microfauna in the soil environment is the major challenge of this plant-microbe technology.
- Sites polluted by complex mixtures of different contaminants may pose challenges for contaminated land practitioners during decision-making.
- Modification of microbe genotypes via genetic engineering could help to overcome the main shortcomings of microbe-assisted phytoremediation and make it more efficient.
- The inability of microbes to grow to sufficient depths to reach subsurface contaminants.
- Sustainable plant-microbe technology depends mainly on the climatic and weather conditions.
- There are also challenges when the contaminated soil is deeper than the rooting zone. This necessitates excavation prior to phytoremediation. Trees have deeper roots that can facilitate remediation at greater depths without excavation. This tree-based technology is referred to as dendroremediation.
- The great challenge of microbe-assisted phytoremediation is the successful use of such inoculum in the field with natural environmental conditions and competition by autochthonous microorganisms.
- Complex and integrated approaches for rhizosphere management are required, since disturbed soils are usually characterized by high complexity and heterogeneity.
- It can be difficult to establish trees in contaminated soils, and once established, they take several years to attain sufficient biomass for efficient rates of phytoremediation.

- The contaminating material should be present within the root zone to be accessible to the root; that is, if metals are tightly bound to the organic portions of the soil, some of the contamination may not be bioavailable to plants.
- Knowledge of soil, contaminant biochemistry, and plant physiology is crucial to assess the likelihood of success at any given site and to troubleshoot if the expected results are not achieved. Choosing appropriate plant and microbial species can also be a challenge.
- The lack of funding from public and private sector agencies for supporting further phytoremediation research is also a major challenge. Seven years on, the government is unable to implement the recommendations due to lack of funds.
- Another important task is to better understand and control the uptake and translocation of organic pollutants in green plants. Numerous pollutants are very hydrophobic. This characteristic and high chemical stability explain why such pollutants are persistent in the environment.
- Knowledge of the choice of suitable plant species and an understanding of soil properties as well as microbial activities that might be lacking are prerequisites for effective phytoremediation.
- A large amount of knowledge is now available on the biochemical processes involved in the detoxification of pollutants inside plant cells. One of the most important challenges is to use this basic scientific information to improve the efficiency of microbe-assisted phytoremediation in the field.
- Another challenge is that there are stressors that affect phytoremediation efficacy of plants in the field that are not encountered in the laboratory or greenhouse conditions. These include variations in temperature, nutrients and precipitation, herbivory (insects and/or animals), plant pathogens, and competition by weed species. Any of these abiotic or biotic stressors can diminish or prevent plant growth in the field, and this will negatively impact the phytoremediation efficacy of plants.
- A challenge frequently encountered in field studies is uneven distribution of contaminants across a site. In laboratory and greenhouse experiments, soils are generally well mixed to achieve a uniform matrix. This may not be possible in the field, even if the site is extensively tilled prior to planting. The spatial heterogeneity of initial contaminant levels results in data scatter, which can make it difficult to statistically show significant treatment effects for field trials.
- Contaminants collected in leaves can be released again to the environment during litter fall.
- Another significant challenge in the demonstration of successful microbe-assisted phytoremediation is the presence of biological organic compounds in soil.
- Degradation of organic contaminants generally requires the concerted action of numerous enzymes, and it is generally impractical to introduce all the genes required for degradation of an organic contaminant into a single plant or its associated microbial genome. It can be difficult to stably

maintain even a single gene in transformed or recombinant organisms, and the desired trait, such as enhanced degradative capacity, is often lost.
- Although genetically engineered microorganisms (GEMs) undoubtedly have greater remediation potential, studies related to their impact on ecosystems and regulation hurdles (related to biosecurity, diversity, end users, government clearance) need to be overcome before commercial use.

Microbe-assisted phytoremediation shows great potential for the cleanup of contaminated sites; however, more comprehensive experiments are needed in order to optimize this method and overcome its limitations. Therefore, various research laboratories are engaged to overcome the limitations of this plant-microbe technology.

12 Conclusions and Future Prospects

Soil and water contamination is a serious worldwide concern; therefore, effective remediation approaches are urgently required. Microbe-assisted phytoremediation is a promising, inexpensive, and eco-friendly rehabilitation approach that uses a broad range of plants and their associated microbes for remediating pollutants present in different environmental matrices. They can metabolize, detoxify, and/or biotransform many refractory pollutants either to obtain carbon and/or energy for their growth or as co-substrates, thus converting them to simpler products such as carbon dioxide (CO_2) or water (H_2O) (Turnau et al. 2005; Phillips et al. 2008; Weyens et al. 2013; Liu et al. 2015). The composition of the microbial community in the soil–root interface can significantly assist plant establishment on heavy metal–contaminated soils by mediating nutrient mineralization and uptake by plants (Wu et al. 2006a,b; Ma et al. 2009a,b). Microbes may reduce the toxicity of metals or increase their bioavailability and thus have some potential to improve phytoextraction efficiency. Different mechanisms have been proposed to explain the beneficial effect of root-associated bacteria, which include production of phytohormones, antioxidant enzymes, biosurfactants, metal-chelating compounds, siderophores, cell wall–degrading enzymes, 1-aminocyclopropane-1-carboxylic acid (ACC) deaminase enzyme, solubilization of minerals, and induced systemic tolerance (Jing et al. 2007; Bonfante and Genre 2010; Liu et al. 2013; Ahemad and Kibret 2014; Cabral et al. 2015; Chandra and Kumar 2015b). The metal-chelating molecules synthesized by bacteria may also act as a key mechanism in heavy metal accumulation and phytoremediation technology. The effects of heavy metals on plant and microbe interactions are complex and may be affected by various factors, such as the chemical nature of metals and concentrations, that should be taken into consideration.

As microbe-assisted phytoremediation is still in its initial stages of research and development, more well-designed and well-documented demonstration projects are necessary to promote phytoremediation as an environmentally friendly and cost-effective technology. The majority of the research has been conducted in laboratories under relatively controlled conditions for short periods of time. To implement microbe-assisted phytoremediation on a larger scale in the environment, more extensive research under field conditions for long durations is required for a better understanding of the potential role of microbes in phytoremediation of environmental pollutants. There are still many areas with poor understanding or scant information where further research is needed. They include:

- In order to implement microbe-assisted phytoremediation in the field, further research is needed to understand the diversity and ecology of plant-associated bacteria in contaminated environments.

- Regulatory acceptability, site assessment, and the potential ecological risks should be specially addressed.
- Researchers must avoid the introduction of invasive plant species as hyperaccumulators that may affect the indigenous floral diversity.
- A major limiting factor for microbe-assisted phytoremediation of recalcitrant organic pollutants is often their low bioavailability. Therefore, there is an urgent need for research aiming at a better understanding of the subtle and complex interactions between pollutants, soil material, plant roots, and microorganisms in the rhizospheric zone.
- Further research is needed to find more efficient hyperaccumulators, which show fast growth, high biomass, high tolerance, and accumulation of metals and other inorganics.
- An extended knowledge of biochemical mechanisms of root exucates, mycorrhizal fungi, and rhizospheric bacteria will contribute to optimize the microbe-assisted phytoremediation process and make it more attractive in the near future.
- The health and phytoremediation efficiency of existing plants could also be improved through the use of conventional breeding techniques and genetic engineering approaches. Both plant physiologists and microbiologists will play a key role in this line of research.
- Little has been done to investigate microorganism-induced changes in the rhizosphere of hyperaccumulator plants in relation to metal accumulation. Similarly, it is difficult to clarify specific features of microbial–plant and microorganism–soil interactions in the rhizosphere.
- There is a need for more pilot and field studies to demonstrate the effectiveness of microbe-assisted phytoremediation technology and increase its acceptance. It is noteworthy that researchers tend to link phytoremediation to biomass production from an economic point of view.
- An additional focus on biomass energy, feedstock for pyrolysis, biofortified products, and carbon sequestration may be necessary for the advancement of research on, and practical applications of, microbe-assisted phytoremediation.
- Multidisciplinary teams of researchers from different backgrounds (e.g., plant physiologists, agronomists, soil scientists, molecular biologists, microbiologists, biotechnologists, chemists, environmental engineers, bioprocess engineers, and government regulators) could accomplish the various tasks for further improvements in microbe-assisted phytoremediation processes.
- The successful use of plant–microbe systems in remediation of environmental pollutants depends on the *in situ* formation of a high number of inoculated bacterial or fungal strains. Thus, the survival and competitive ability of the microbes must be evaluated under realistic field conditions.
- Further research is needed in order to understand the plant–microbe interactions during removal of contaminants, including micropollutants, in different types of treatment wetlands such as constructed wetlands and floating treatment wetlands. Based on this knowledge, treatment of wetlands design and operational parameters could be improved to achieve more efficient pollutant removal from contaminated water or wastewater.

- Microbe-assisted phytoremediation efficiency is also limited by the climatic and geological conditions of the mined site to be decontaminated and requires highly technical and expert project designers with plenty of field experience that carefully choose the proper species and manage the entire system to maximize the efficiency of phytostabilization.
- Rhizobacteria encounter soil solution before it enters the root, and the sequestration of heavy metals by rhizobacteria from soil solution may play an important part in plant metal uptake. The role played by bacteria from soil solution in plant metal uptake is still poorly understood.
- To combine the characteristics of many potential genes, transgenic plants may also provide a good option in the future for microbe-assisted phytoremediation.
- Since the biotic/abiotic stress in multi-metal–polluted field soils greatly influences the activity, composition, and function of the inoculated microbes, microbial-mediated processes may be dependent on the nutrient composition and properties of rhizosphere soils. Moreover, various stimuli in the rhizosphere could be also associated with production of metabolites (e.g., siderophores, organic acids), including nutrient deficiency (P, Fe) and exposure to toxic metals. Thus, characterizing the physico-chemical-biological features of target contaminated soils may be important for making microbe-assisted phytoremediation processes successful.
- More demonstration projects are also urgently required to provide recommendations and convince regulators, decision-makers, and the general public of the applicability of a green approach for the treatment of soils, groundwater, and wastewater contaminated by toxic metals, organic pollutants, and radionuclides.
- The isolation of various plant-associated microbes and characterization of their beneficial metabolites/processes are time consuming since they require the analysis of more than thousands of isolates. Thus, a strong molecular research effort is required in order to find specific biomarkers associated with the beneficial microbes for efficient phytoremediation.
- Basic research is still lacking in order to efficiently exploit the immense possibilities offered by plant-microbes technologies. In this regard, the integration of new molecular tools with previous knowledge on the genetics, physiology, and biochemistry of plants is expected to significantly advance our understanding of the relevant mechanisms for pollutant degradation. This information will be used to create superior varieties via genetic engineering, an approach that has already proven feasible.
- Microbe-assisted phytoremediation is strongly not only influenced by soil characteristics but also substantially influences microbial colonization and activity. Thus, the importance of such parameters should be considered when designing phytoremediation applications.
- An improved fundamental knowledge of the physiological traits of rhizosphere microorganisms and their impact on rhizosphere processes, which are especially relevant for the remediation of disturbed soils, will be

essential to allow an increased and successful use of microbial inoculum in the field.

- To further advance our knowledge, microbe-assisted phytoremediation research requires more collaborative studies involving expertise from different fields such as botany, plant physiology, biochemistry, geochemistry, agricultural engineering, microbiology, and genetic engineering, among others. In the years to come, as in other areas, plant genetic engineering for improved phytoremediation could also benefit from the data of genomic and postgenomic projects, including proteomics.

- Analysis of sequenced genomes, the characterization of yet-unknown genes, and the identification of genes expressed during plant root colonization will help to improve our understanding of the colonization process and the interaction of beneficial microbes with plants.

- Complete genome sequences for several environmentally relevant microorganisms, mechanisms of microbial chelators-metal complex uptake in plants, factors influencing the solubility and plant availability of nutrients/heavy metals, signaling processes that occur between plant roots and microbes: these types of analysis will surely prove useful for exploring the mechanism of metal–microbe–plant interactions. Moreover, such knowledge may enable us to improve the performance and use of beneficial microbes as inoculants for microbial-assisted phytoremediation application at field scale.

- Most results of endophyte-assisted phytoremediation were obtained from *in vitro* studies, but the phytoremediation process may be influenced by various environmental factors, and more work should be carried out to document the role of naturally adapted indigenous endophytes.

- More studies should be conducted to understand the diversity and ecology of endophytes/plant growth-promoting rhizobacteria (PGPR) associated with the plants growing in multiple metal/organic-contaminated soils, as the interaction between phytoremediation-enhancing endophytes/rhizobacteria may be more important for endophyte/rhizobacteria-assisted phytoremediation at the field level.

- In depth research is needed to investigate mechanisms involved in mobilization, degradation, various chemical aspects of metal accumulation, and transfer of metals to develop future strategies and optimize phytoextraction by microbe–plant systems. In addition, further studies are needed to assess the robustness of the microbial community for efficient bioremediation of heavy metals in natural field conditions. A more comprehensive understanding of these microbes in their natural environment is important for phytoremediation technology to reach its full potential.

- Although plants and their associated endophytes/PGPR can degrade a wide range of contaminants, many compounds are degraded slowly or are not degraded at all. In order to help this technology to be rapidly adopted and put into widespread use, more engineered endophytes/PGPR with appropriate biodegradative capabilities should be produced.

- The mechanisms involved in endophyte-assisted phytoremediation still require further evaluation, and more investigations are needed to better realize the potential of these fascinating endophytes for improving the soil environment.

- Since plant heavy metal tolerance and uptake depend on various factors, including the bacterium, soil composition, metal type, and concentration, the scientific community should strongly focus on the selection of the most suitable plants to overcome such environmental constraints.
- We need to further understand the mechanisms involved in mobilization and transfer of metals in order to develop future strategies and optimize the phytoextraction process. Such knowledge may enable us to understand the role and mechanism of soil rhizobacteria on phytoremediation.
- In metal phytoremediation studies, different bacterial strains affect plant growth and metal uptake in different ways. Thus, further research should be focused on how a plant's metal tolerance is affected by its associated bacteria.
- Concentrated efforts are needed to explore the beneficial traits of pollutant-degrading endophytic bacteria to get the maximum outcome of the combined use of plants and endophytic bacteria in the field.
- The use of genetically engineered endophytic bacteria carrying additional genes for improved colonization, interaction with the plant, or degradation may be a more promising strategy in improving phytoremediation of soil and water polluted with organic compounds. However, in the case of genetic engineering, an assessment of potential side effects should be performed.
- To maximize the outcomes of plant–endophyte interactions, high-throughput strategies are needed in which all or most of the genes involved in organic pollutant degradation, proteins, or even metabolites in an organism are subjected to functional analysis.
- It is expected that the interactions between plants and endophytic bacteria will be explored more efficiently by using the latest molecular biology techniques, and endophyte-assisted phytoremediation will be effectively applied in the field for the cleanup of soil and water polluted with organic compounds.
- Previous work has mainly focused on endophytic bacteria and their assistance in phytoremediation, but little is known about endophytic fungi. More work should be conducted on endophytic fungi with metal resistance and/or organic-contaminant-degrading capability and their ability to assist in phytoremediation.
- Recent plant biotechnology approaches involving the introduction of specialized bacterial endophytes in plants or the design of genetically engineered plants containing interesting bacterial genes create new perspectives for future phytoremediation protocols.
- The molecular engineering of both microbes and plants with desired genes would help immensely to enhance the efficiency of growth-promoting rhizobacteria-mediated or plant-based remediation of contaminated sites.
- Not much data are yet available on the field performance of transgenic microorganisms associated with plants in phytoremediation. Established field trials are, therefore, urgently needed to make it a commercially viable and acceptable technology.

- Most studies on endophytes and rhizobacteria are commonly based on experimental manipulations and are rarely based on variable, field-realistic conditions. Future research should move beyond these limitations and identify how plant–endophyte and rhizobacteria partnerships are working in tandem under different scenarios.
- To employ microbe-assisted phytoremediation of either organic, metal or metal-organic contaminants on a large scale in the environment, it will be necessary to convince regulatory bodies in various jurisdictions that the deliberate release of selected, or even engineered, microbes to the environment should be viewed not only as benign, but in fact as beneficial.
- Understanding the mechanistic basis of the physical, chemical, and biological rhizosphere processes and the interactions between hyperaccumulators and non-accumulators and plant growth-promoting rhizobacteria and/or endophytic bacteria will be important in modeling better the full impact of phytoremediation in the restoration of derelict lands.
- For phytoextraction of heavy metals, use of non-food crops such as those used in timber or other commercial purposes (not involving human or animal consumption) can be targeted. This will result in removal of heavy metals from the soil and non-transfer to the food chain.
- The selection and inoculation of plants with indigenous fungal stains, rather than the application of allochtonous inocula, should be recommended to get better effects in phytoremediation programs.

Although this approach has already been found to be effective in the field, it is likely that during the next five to ten years, its use will become widespread. That is not to say that no technical difficulties remain; however, the problems appear to be tractable. For widespread future use of this technique, public awareness of this technology is necessary to enhance its acceptability as global sustainable technology.

References

Abbaslou H, Bakhtiari S, Hashemi SS. 2018. Rehabilitation of iron ore mine soil contaminated with heavy metals using rosemary phytoremediation-assisted mycorrhizal arbuscular fungi bioaugmentation and fibrous clay mineral immobilization. *Iranian Journal of Science and Technology, Transactions A: Science* 42(2):431–444.

Abedinzadeh M, Etesami H, Alikhani HA. 2019. Characterization of rhizosphere and endophytic bacteria from roots of maize (*Zea mays* L.) plant irrigated with wastewater with biotechnological potential in agriculture. *Biotechnology Reports (Amst)* 21:e00305.

Abhilash PC, Powell JR, Singh HB, Singh BK. 2012. Plant–microbe interactions: Novel applications for exploitation in multipurpose remediation technologies. *Trends in Biotechnology* 30(8):416–420.

Abhilash PC, Singh B, Srivastava P, Schaeffer A, Singh N. 2013. Remediation of lindane by *Jatropha curcas* L: Utilization of multipurpose species for rhizoremediation. *Biomass and Bioenergy* 51:189–193.

Aboudrar W, Schwartz C, Morel JL, Boularbah A. 2013. Effect of nickel-resistant rhizosphere bacteria on the uptake of nickel by the hyperaccumulator *Noccaea caerulescens* under controlled conditions. *Journal of Soils and Sediments* 13(3):501–507.

Aboughalma H, Bi R, Schlaak M. 2008. Electrokinetic enhancement on phytoremediation in Zn, Pb, Cu and Cd contaminated soil using potato plants. *Journal of Environmental Science and Health, Part A Toxic/Hazardous Substances and Environmental Engineering* 43(8):926–933.

Abou-Shanab RAI, Angle JS, Chaney RL. 2006. Bacterial inoculants affecting nickel uptake by *Alyssum murale* from low, moderate and high Ni soils. *Soil Biology and Biochemistry* 38(9):2882–2889.

Abou-Shanab RI, Angle JS, Delorme TA, Chaney RL, van Berkum P, Moawad H, Ghanem K, Ghozlan HA. 2003. Rhizobacterial effects on nickel extraction from soil and uptake by *Alyssum murale*. *New Phytologist* 158:219–224.

Abu-Elsaoud AM, Nafady NA, Abdel-Azeem AM. 2017. Arbuscular mycorrhizal strategy for zinc mycoremediation and diminished translocation to shoots and grains in wheat. *PLOS ONE* 12(11):e0188220.

Adediran GA, Ngwenya BT, Mosselmans JFW, Heal KV, Harvie BA. 2015. Mechanism behind bacteria induced plant growth promotion and Zn accumulation in *Brassica juncea*. *Journal of Hazardous Materials* 283:490–499.

Adeola FO. 2004. Boon or Bane? The environmental and health impacts of persistent organic pollutants (POPs). *Human Ecology Review* 11(1):27–35.

Afzal M, Rehman K, Shabir G, Tahseen R, Ijaz A, Hashmat AJ, Brix H. 2019. Large-scale remediation of oil-contaminated water using floating treatment wetlands. *NPJ Clean Water* 2. Article 3.

Afzal M, Yousaf S, Reichenauer TG, Kuffner M, Sessitsch A. 2011. Soil type affects plant colonization, activity and catabolic gene expression of inoculated bacterial strains during phytoremediation of diesel. *Journal of Hazardous Materials* 186:1568–1575.

Afzal M, Yousaf S, Reichenauer TG, Sessitsch A. 2012. The inoculation method affects colonization and performance of bacterial inoculant strains in the phytoremediation of soil contaminated with diesel oil. *International Journal of Phytoremediation* 14:35–47.

Ahemad M. 2015. Phosphate-solubilizing bacteria-assisted phytoremediation of metalliferous soils: A review. *3 Biotech* 5:111–121.

Ahemad M. 2019. Remediation of metalliferous soils through the heavy metal resistant plant growth promoting bacteria: Paradigms and prospects. *Arabian Journal of Chemistry* 12(7):1365–1377.

Ahemad M, Kibret M. 2014. Mechanisms and applications of plant growth promoting rhizobacteria: Current perspective. *Journal of King Saud University—Science* 26(1):1–20.

Ahkami AH, Allen White III R, Handakumburaa PP, Jansson C. 2017. Rhizosphere engineering: Enhancing sustainable plant ecosystem productivity. *Rhizosphere* 3(Part 2):233–243.

Ahmed E, Holmström SJM. 2014. Siderophores in environmental research: Roles and applications. *Microbial Biotechnology* 7(3):196–208.

Akhter K, Ghous T, Andleeb S et al. 2017. Bioaccumulation of heavy metals by metal-resistant bacteria isolated from tagetes minuta rhizosphere, growing in soil adjoining automobile workshops. *Pakistan Journal of Zoology* 49(5):1841–1846.

Alarcon A, Davies Jr, FT, Autenrieth RL, Zuberer DA. 2008. Arbuscular mycorrhiza and petroleum-degrading microorganisms enhance phytoremediation of petroleum-contaminated soil. *International Journal of Phytoremediation* 10:251–263.

Al-Awadhi H, El-Nemr I, Mahmoud H, Sorkhoh NA, Radwan SS. 2009. Plant-associated bacteria as tools for the phytoremediation of oily nitrogen-poor soils. *International Journal of Phytoremediation* 11:11–27.

Al-Enazy AAR, Al-Oud SS, Al-Barakah FN, Usman AR. 2017. Role of microbial inoculation and industrial by-product phosphogypsum in growth and nutrient uptake of maize (*Zea mays* L.) grown in calcareous soil. *Journal of Science, Food and Agriculture* 97(11):3665–3674.

Ali A, Guo D, Mahar A, Ma F, Li R, Shen F, Wang P, Zhang Z. 2017. *Streptomyces pactum* assisted phytoremediation in Zn/Pb smelter contaminated soil of Feng County and its impact on enzymatic activities. *Scientific Reports* 7:46087.

Ali H, Khan E, Ilahi I. 2019. Environmental chemistry and ecotoxicology of hazardous heavy metals: Environmental persistence, toxicity, and bioaccumulation. *Journal of Chemistry* 2019:14. Article ID 6730305.

Ali H, Khan E, Sajadc MA. 2013. Phytoremediation of heavy metals—Concepts and applications. *Chemosphere* 91(7):869–881.

Alori ET, Glick BR, Babalola OO. 2017. Microbial phosphorus solubilization and its potential for use in sustainable agriculture. *Frontiers in Microbiology* 8:971.

Altomare C, Norvell WA, Björkman T, Harman GE. 1999. Solubilization of phosphates and micronutrients by the plant-growth-promoting and biocontrol fungus *Trichoderma harzianum* Rifai. *Applied and Environmental Microbiology* 65(7):2926–2933.

Andreolli M, Lampis S, Poli M, Gullner G, Biróc B, Vallini G. 2013. Endophytic *Burkholderia fungorum* DBT1 can improve phytoremediation efficiency of polycyclic aromatic hydrocarbons. *Chemosphere* 92:688–694.

Andria V, Reichenauer TG, Sessitsch A. 2009. Expression of alkane monooxygenase (alkB) genes by plant-associated bacteria in the rhizosphere and endosphere of Italian ryegrass (*Lolium multiflorum* L.) grown in diesel contaminated soil. *Environmental Pollution* 157:3347–3350.

Ankohina T, Kochetkov V, Zelenkova N, Balakshina V, Boronin A. 2004. Biodegradation of phenanthrene by *Pseudomonas* bacteria bearing rhizospheric plasmids in model plant-microbial associations. *Applied Biochemistry and Microbiology* 40:568–572.

Antoun H, Kloepper JW. 2001. Plant growth promoting rhizobacteria. In: Brenner S, Miller J (eds) *Encyclopedia of Genetics*. New York: Academic, pp. 1477–1480.

Antoun H, Prevost D. 2005. Ecology of plant growth promoting rhizobacteria. In: Siddiqui ZA (ed.) *PGPR: Biocontrol and Biofertilization*. Dordrecht: Springer, pp. 1–38.

Arkhipova TN, Prinsen E, Veselov SU, Martinenko EV, Melentiev AI, Kudoyarova GR. 2007. Cytokinin producing bacteria enhance plant growth in drying soil. *Plant and Soil* 292:305.

Arora K, Sharma S, Monti A. 2016. Bio-remediation of Pb and Cd polluted soils by switch grass: A case study in India. *International Journal of Phytoremediation* 18:704–709.

Arslan M, Imran A, Khan QM, Afzal M. 2017. Plant–bacteria partnerships for the remediation of persistent organic pollutants. *Environmental Science and Pollution Research* 24:4322–4336.

Ashraf S, Afzal M, Naveed M, Shahid M, Zahir ZA. 2018a. Endophytic bacteria enhance remediation of tannery effluent in constructed wetlands vegetated with *Leptochloa fusca. International Journal of Phytoremediation* 20(2):121–128.

Ashraf S, Afzal M, Rahman K, Naveed M, Zahir ZA. 2018b. Plant–endophyte synergism in constructed wetlands enhances the remediation of tannery effluent. *Water Science and Technology* 77:1262–1270.

Ashraf S, Ali Q, Zahir ZA, Ashraf S, Asghar HN. 2019. Phytoremediation: Environmentally sustainable way for reclamation of heavy metal polluted soils. *Ecotoxicology and Environmental Safety* 174:714–727.

Audet P, Charest C. 2006. Effects of AM colonization on "wild tobacco" plants grown in zinc-contaminated soil. *Mycorrhiza* 16(4):277–283.

Azubuike CC, Chikere CB, Okpokwasili GC. 2016. Bioremediation techniques–classification based on site of application: Principles, advantages, limitations and prospects. *World Journal of Microbiology and Biotechnology* 32:180.

Babu AG, Kim JD, Oh BT. 2013. Enhancement of heavy metal phytoremediation by *Alnus firma* with endophytic *Bacillus thuringiensis* GDB-1. *Journal of Hazardous Materials* 250–251:477–483.

Baker AJM. 1981. Accumulator and excluder-strategies in the response of plant to heavy metals. *Journal of Plant Nutrition* 3:643–654.

Baker AJM, Brooks RR. 1989. Terrestrial higher plants which hyperaccumulate metallic elements: A review of their distribution, ecology and phytochemistry. *Biorecovery* 1:81–126.

Baker AJM, Reeves RD, Hajar ASM. 1994. Heavy metal accumulation and tolerance in British populations of the metallophyte *Thlaspi caerulescens* J. et C. Presl. (*Brassicaceae*). *New Phytologist* 127:61–68.

Baker AJM, Walker PL. 1989. Ecophysiology of metal uptake by tolerant plants. In: Shaw AJ (ed.) *Heavy Metal Tolerance in Plants: Evolutionary Aspects.* Boca Raton, FL: CRC Press, pp. 155–177.

Bakker PA, Berendsen RL, Doornbos RF, Wintermans PC, Pieterse CM. 2013.The rhizosphere revisited: Root microbiomics. *Frontiers in Plant Science* 30(4):165.

Barac T, Taghavi S, Borremans B, Provoost A, Oeyen L, Colpaert JV, Vangronsveld J, van der Lelie D. 2004. Engineered endophytic bacteria improve phytoremediation of water-soluble, volatile, organic pollutants. *Nature Biotechnology* 22(5):583–588.

Barkay T, Wagner-Dobler I. 2005. Microbial transformations of mercury: Potentials, challenges, and achievements in controlling mercury toxicity in the environment. *Advances in Applied Microbiology* 57:1–52.

Barzanti R, Ozino F, Bazzicalupo M, Gabbrielli R, Galardi F, Gonnelli C, Mengoni A. 2007. Isolation and characterization of endophytic bacteria from the nickel hyperaccumulator plant *Alyssum bertolonii. Microbial Ecology* 53(2):306–316.

Bashan Y, Holguin G, de-Bashan LE. 2004. Azospirillum–plant relationships: Physiological, molecular, agricultural, and environmental advances (1997–2003). *Canadian Journal of Microbiology* 50:521–577.

Bashan Y, Kamnev AA, de-Bashan LE. 2013. Tricalcium phosphate is inappropriate as a universal selection factor for isolating and testing phosphate-solubilizing bacteria that enhance plant growth: A proposal for an alternative procedure. *Biology and Fertility of Soils* 49(4):465–479.

Beattie GA. 2006. Plant-associated bacteria: Survey, molecular phylogeny, genomics and recent advances. In: Gnanamanickam SS (ed.) *Plant-Associated Bacteria.* Dordrecht: Springer, pp. 1–56.

Becerra-Castro C, Monterroso C, Garcia-Leston M, Prieto-Fernandez A, Acea MJ, Kidd PS. 2009. Rhizosphere microbial densities and trace element tolerance of the nickel hyperaccumulator *Alyssum serpyllifolium* subsp. lusitanicum. *International Journal of Phytoremediation* 11:525–541.

Becerra-Castro C, Prieto-Fernández Á, Kidd PS et al. 2013. Improving performance of *Cytisus striatus* on substrates contaminated with hexachlorocyclohexane (HCH) isomers using bacterial inoculants: Developing a phytoremediation strategy. *Plant and Soil* 362:247–260.

Begum N, Qin C, Ahanger MA, Raza S, Khan MI, Ashraf M, Ahmed N, Zhang L. 2019. Role of arbuscular mycorrhizal fungi in plant growth regulation: Implications in abiotic stress tolerance. *Frontiers in Plant Science* 10:1068.

Beijerinck MW. 1901. Uber oligonitrophile Mikroben, *Zbl. Backt.* 7:561–582.

Beneduzi A, Ambrosini A, Passaglia LM. 2012. Plant growth-promoting rhizobacteria (PGPR): Their potential as antagonists and biocontrol agents. *Genetics and Molecular Biology* 35(4):1044–1051.

Benyó D, Horváth E, Németh E et al. 2016. Physiological and molecular responses to heavy metal stresses suggest different detoxification mechanism of *Populus deltoides* and *P. × canadensis*. *Journal of Plant Physiology* 201:62–70.

Bhargavi VLN, Sudha PN. 2015. Removal of heavy metal ions from soil by electrokinetic assisted phytoremediation method. *International Journal of Chemical Technology and Research* 8:192–202.

Bissonnette L, St-Arnaud M, Labrecque M. 2010. Phytoextraction of heavy metals by two *Salicaceae* clones in symbiosis with arbuscular mycorrhizal fungi during the second year of a field trial. *Plant and Soil* 332:55–67.

Bolan N, Kunhikrishnan A, Thangarajan R, Kumpiene J, Park J, Makino T, Kirkham MB, Scheckel K. 2014. Remediation of heavy metal(loid)s contaminated soils--to mobilize or to immobilize? *Journal of Hazardous Materials* 266:141–166.

Böltner D, Godoy P, Munoz-Rojas J, Duque E, Moreno-Morillas S, Sánchez L, Ramos JL. 2008. Rhizoremediation of lindane by root colonizing *Sphingomonas*. *Microbiology and Biotechnology* 1:87–93.

Bonfante P, Genre A. 2010. Mechanisms underlying beneficial plant—fungus interactions in mycorrhizal symbiosis. *Nature Communications* 1:48.

Bouffaud ML, Renoud S, Moënne-Loccoz Y, Muller D. 2016. Is plant evolutionary history is plant evolutionary history impacting recruitment of diazotrophs and *nifH* expression in the rhizosphere? *Scientific Reports* 6:21690.

Braga RM, Dourado MN, Araújo WL. 2016. Microbial interactions: Ecology in a molecular perspective. *Brazilian Journal of Microbiology* 47(Suppl 1):86–98.

Brandstatter I, Kieber JJ. 1998. Two genes with similarity to bacterial response regulators are rapidly and specifically induced by cytokinin in Arabidopsis. *Plant Cell* 10(6):1009–1019.

Brazil GM, Kenefick L, Callanan M et al. 1995. Construction of a rhizosphere pseudomonad with potential to degrade polychlorinated biphenyls and detection of bph gene expression in the rhizosphere. *Applied and Environmental Microbiology* 61:1946–1952.

Brix H. 1994. Functions of macrophytes in constructed wetlands. *Water Science and Technology* 29(4):71–78.

Brix H. 1997. Do macrophytes play a role in constructed treatment wetlands? *Water Science and Technology* 35(5):11–17.

Brix H, Schierup HH. 1989. The use of aquatic macrophytes in water-pollution control. *Ambio* 18:100–107.

Burd GI, Dixon DG, Glick BR. 1998. A plant growth-promoting bacterium that decreases nickel toxicity in plant seedlings. *Applied and Environmental Microbiology* 64:3663–3668.

Burd GI, Dixon DG, Glick BR. 2000. Plant growth-promoting bacteria that decrease heavy metal toxicity in plants. *Canadian Journal of Microbiology* 46:237–245.

Bürgmann H, Widmer F, Von Sigler W, Zeyer J. 2004. New molecular screening tools for analysis of free-living diazotrophs in soil. *Applied and Environmental Microbiology* 70(1):240–247.

Burken JG, Schnoor JL. 1997. Uptake and metabolism of atrazine by poplar trees. *Environmental Science and Technology* 31:1399–1406.

Buysens S, Heungens K, Poppe J, Hofte M. 1996. Involvement of pyochelin and pyoverdin in suppression of pythium-induced damping-off of tomato by *Pseudomonas aeruginosa* 7NSK2. *Applied and Environmental Microbiology* 62(3):865–871.

Caballero-Mellado J, Onofre-Lemus J, Estrada-de los Santos P, Martınez-Aguilar L. 2007. The tomato rhizosphere, an environment rich in nitrogen-fixing *Burkholderia* species with capabilities of interest for agriculture and bioremediation. *Applied and Environmental Microbiology* 73(16):5308–5319.

Cabral L, Soares CR, Giachini AJ, Siqueira JO. 2015. Arbuscular mycorrhizal fungi in phytoremediation of contaminated areas by trace elements: Mechanisms and major benefits of their applications. *World Journal of Microbiology and Biotechnology* 31(11):1655–1664.

Calheiros CS, Rangel AO, Castro PM. 2007. Constructed wetland systems vegetated with different plants applied to the treatment of tannery wastewater. *Water Research* 41:1790–1798.

Calheiros CSC, Duque AF, Moura A, Henriques IS, Correia A, Rangel AOSS, Castro PML. 2009. Changes in the bacterial community structure in two-stage constructed wetlands with different plants for industrial wastewater treatment. *Bioresource Technology* 100:3228–3235.

Cameselle C, Chirakkara RA, Reddy KR. 2013. Electrokinetic-enhanced phytoremediation of soils: Status and opportunities. *Chemosphere* 93(4):626–636.

Cang L, Wang Q–Y, Zhou D–M, Xu H. 2011. Effects of electrokinetic-assisted phytoremediation of a multiple-metal contaminated soil on soil metal bioavailability and uptake by Indian mustard. *Separation and Purification Technology* 79(2):246–253.

Carnevali O, Santangeli S, Forner-Piquer I, Basili D, Maradonna F. 2018. Endocrine-disrupting chemicals in aquatic environment: What are the risks for fish gametes?. *Fish Physiology and Biochemistry* 44(6):1561–1576.

Carpenter DO. 2011. Health effects of persistent organic pollutants: The challenge for the Pacific Basin and for the world. *Reviews in Environment and Health* 26(1):61–69.

Carvalho PN, Basto MC, Almeida CM, Brix H. 2014. A review of plant-pharmaceutical interactions: From uptake and effects in crop plants to phytoremediation in constructed wetlands. *Environmental Science and Pollution Research International* 21(20):11729–11763.

Chandra R, Kumar V, Singh K. 2018a. Hyperaccumulator versus nonhyperaccumulator plants for environmental waste management. In: Chandra R, Dubey NK, Kumar V (eds) *Phytoremediation of Environmental Pollutants*. Boca Raton: CRC Press.

Chandra R, Kumar V, Tripathi S, Sharma P. 2018b. Heavy metal phytoextraction potential of native weeds and grasses from endocrine-disrupting chemicals rich complex distillery sludge and their histological observations during in situ phytoremediation. *Ecological Engineering* 111:143–156.

Chandra R, Kumar V. 2015a. Biotransformation and biodegradation of organophosphates and organohalides. In: Chandra R (ed.) *Environmental Waste Management*. Boca Raton: CRC Press.

Chandra R, Kumar V. 2015b. Mechanism of wetland plant rhizosphere bacteria for bioremediation of pollutants in an aquatic ecosystem. In: Chandra R (ed.) *Advances in Biodegradation and Bioremediation of Industrial Waste*. Boca Raton: CRC Press.

Chandra R, Kumar V. 2017a. Detection of androgenic-mutagenic compounds and potential autochthonous bacterial communities during *in situ* bioremediation of post methanated distillery sludge. *Frontiers in Microbiology* 8:887.

Chandra R, Kumar V. 2017b. Detection of *Bacillus* and *Stenotrophomonas* species growing in an organic acid and endocrine-disrupting chemicals rich environment of distillery spent wash and its phytotoxicity. *Environmental Monitoring and Assessment* 189:26.

Chandra R, Kumar V. 2017c. Phytoextraction of heavy metals by potential native plants and their microscopic observation of root growing on stabilized distillery sludge as a prospective tool for in-situ phytoremediation of industrial waste. *Environmental Science and Pollution Research* 24:2605–2619.

Chandra R, Kumar V. 2018. Phytoremediation: A green sustainable technology for industrial waste management. In: Chandra R, Dubey NK, Kumar V (eds) *Phytoremediation of Environmental Pollutants*. Boca Raton: CRC Press.

Chandra R, Saxena G, Kumar V. 2015. Phytoremediation of environmental pollutants: An eco-sustainable green technology to environmental management. In: Chandra R (ed.) *Advances in Biodegradation and Bioremediation of Industrial Waste*. Boca Raton: CRC Press, pp. 1–29.

Chaney RL. 1983. Plant uptake of inorganic waste constitutes. In: Parr JF, Marsh PB, Kla JM (eds) *Land Treatment of Hazardous Wastes*. Park Ridge, NJ: Noyes Data Corp., pp. 50–76.

Chen B, Nayuki K, Kuga Y, Zhang X, Wu S, Ohtomo R. 2018b. Uptake and intraradical immobilization of cadmium by arbuscular mycorrhizal fungi as revealed by a stable isotope tracer and synchrotron radiation μX-Ray fluorescence analysis. *Microbes and Environment* 33(3):257–263.

Chen B, Shen H, Li X et al. 2004a. Effects of EDTA application and arbuscular mycorrhizal colonization on growth and zinc uptake by maize (*Zea mays* L.) in soil experimentally contaminated with zinc. *Plant and Soil* 261:219.

Chen B, Shen H, Li X, Feng G, Christie P. 2004b. Effects of EDTA application and arbuscular mycorrhizal colonization on growth and zinc uptake by maize (*Zea mays* L.) in soil experimentally contaminated with zinc. *Plant and Soil* 26:1219–1229.

Chen BD, Zhu YG, Smith FA. 2006a. Effects of arbuscular mycorrhizal inoculation on uranium and arsenic accumulation by Chinese brake fern (*Pteris vittata* L.) from a uranium mining-impacted soil. *Chemosphere* 62(9):1464–1473.

Chen L, Luo S, Li X, Wan Y, Chen J, Liu C. 2014. Interaction of Cd-hyperaccumulator *Solanum nigrum* L. and functional endophyte *Pseudomonas* sp. Lk9 on soil heavy metals uptake. *Soil Biology and Biochemistry* 68:300–308.

Chen M, Arato M, Borghi L, Nouri E, Reinhardt D. 2018a. Beneficial services of arbuscular mycorrhizal fungi – from ecology to application. *Frontiers in Plant Science* 9:1270.

Chen YP, Rekha PD, Arun AB, Shen FT, Lai WA, Young CC. 2006b. Phosphate solubilizing bacteria from subtropical soil and their tricalcium phosphate solubilizing abilities. *Applied Soil Ecology* 34(1):33–41.

Chen Z, Cuervo DP, Müller JA, Wiessner A, Köser H, Vymazal J, Kuschk P. 2016. Hydroponic root mats for wastewater treatment. *Environmental Science and Pollution Research* 23:15911–15928.

Chen Z, Ma S, Liu LL. 2008. Studies on phosphorus solubilizing activity of a strain of phosphobacteria isolated from chestnut type soil in China. *Bioresource Technology* 99:6702–6707.

Chern EC, Tsai DW, Ogunseitan OA. 2007. Deposition of glomalin-related soil protein and sequestered toxic metals into watersheds. *Environmental Science & Technology* 41(10):3566–3572.

Chilingar GV, Lolf WW, Khilyuk LF, Katz SA. 1997. Electrobioremediation of soils contaminated with hydrocarbons and metals: Progress report. *Energy Sources* 19(2):129–146.

Cobbett C, Goldsbrough P. 2002. Phytochelatins and metallothioneins: Roles in heavy metal detoxification and homeostasis. *Annual Review of Plant Biology* 53:159–182.

Compant S, Clément C, Sessitsch A. 2010. Plant growth-promoting bacteria in the rhizo- and endosphere of plants: Their role, colonization, mechanisms involved and prospects for utilization. *Soil Biology and Biochemistry* 42(5):669–678

Compant S, Reiter B, Sessitsch A, Nowak J, Clément C, Ait Barka E. 2005. Endophytic colonization of Vitis vinifera L. by plant growth-promoting bacterium *Burkholderia* sp. strain PsJN. *Applied and Environmental Microbiology* 71(4):1685–1693.

Conrath U, Beckers GJM, Flors V et al. . Priming: Getting ready for battle. *Molecular Plant Microbe Interaction* 19:1062–1071.

Cunningham SD, Berti WR. 1993. Remediation of contaminated soils with green plants: An overview. *In Vitro Cellular & Developmental Biology—Plant* 29(4):207–212.

Cunningham SD, Berti WR, Huang JW. 1995. Phytoremediation of contaminated soils. *Trends in Biotechnology* 13:393–397.

Cunningham SD, Shann JR, Crowley DE, Anderson TA. 1997. Phytoremediation of contaminated water and soil. In: Kruger EL, Anderson TA, Coats JR (eds) *Phytoremediation of Soil and Water Contaminants, ACS Symposium Series 664.* Washington, DC: American Chemical Society, pp. 2–19.

Dary M, Chamber-Pérez MA, Palomares AJ, Pajuelo E. 2010. *"In situ"* phytostabilisation of heavy metal polluted soils using *Lupinus luteus* inoculated with metal resistant plant-growth promoting rhizobacteria. *Journal of Hazardous Materials* 177:323–330.

Davey ME, O'toole GA. 2000. Microbial biofilms: From ecology to molecular genetics. *Microbiology And Molecular Biology Reviews* 64(4):847–867.

Deikman J. 1997. Molecular mechanisms of ethylene regulation of gene transcription. *Physiologia Plantarum* 100(3):561–566

De Jong J. 1976. The purification of wastewater with the aid of rush or reed ponds. In: Tourbier J, Pierson JRW (eds) *Biological Control of Water Pollution Pollution.* Philadelphia: Pennsylvania University Press, pp. 133–139.

de la Providencia IE, Stefani FO, Labridy M, St-Arnaud M, Hijri M. 2015. Arbuscular mycorrhizal fungal diversity associated with Eleocharis obtusa and Panicum capillare growing in an extreme petroleum hydrocarbon-polluted sedimentation basin. *FEMS Microbiol Letters* 362(12):fnv081.

Dennis PG, Miller AJ, Hirsch PR. 2010. Are root exudates more important than other sources of rhizodeposits in structuring rhizosphere bacterial communities? *FEMS Microbiology Ecology* 72(3):313–327.

de Souza MO, Chu D, Zhao M, Zayed AM, Ruzin SE, Schichnes D, Terry N. 1999. Rhizosphere bacteria enhance selenium accumulation and volatilization by Indian Mustard. *Plant Physiology* 199:565–573.

Dickinson NM, Baker AJM, Doronila A, Laidlaw S, Reeves RD. 2009. Phytoremediation of inorganics: Realism and synergies. *International Journal of Phytoremediation* 11(2):97–114.

Dimkpa C, Weinand T, Asch F. 2009. Plant–rhizobacteria interactions alleviate abiotic stress conditions. *Plant, Cell & Environment* 32(12):1682–1694.

Dimkpa CO, Svatos A, Merten D, Buchel G, Kothe E. 2008. Hydroxamate siderophores produced by *Streptomyces acidiscabies* E13 bind nickel and promote growth in cowpea (*Vigna unguiculata* L.) under nickel stress. *Canadian Journal of Microbiology* 54(3):163–172.

Dobrowolski R, Szcześ A, Czemierska M, Jarosz-Wikołazka A. 2017. Studies of cadmium(II), lead(II), nickel(II), cobalt(II) and chromium(VI) sorption on extracellular polymeric substances produced by *Rhodococcus opacus* and *Rhodococcus rhodochrous*. *Bioresource Technology* 225:113–120.

Dotaniya ML, Rajendiran S, Dotaniya CK et al. 2018. Microbial Assisted Phytoremediation for heavy metal contaminated soils. In: Kumar V, Kumar M, Prasad R (eds) *Phytobiont and Ecosystem Restitution*. Singapore: Springer. https://doi.org/10.1007/978-981-13-1187-1_16

Doty SL. 2008. Enhancing phytoremediation through the use of transgenics and endophytes. *New Phytologist* 179:318–333.

Doty SL, Dosher MR, Singleton GL et al. 2005. Identification of an endophytic *Rhizobium* in stems of Populus. *Symbiosis* 39:27–35.

El-Deeb B, Gherbawy Y, Hassan S. 2012. Molecular characterization of endophytic bacteria from metal hyperaccumulator aquatic plant (*Eichhornia crassipes*) and its role in heavy metal removal. *Geomicrobiology Journal* 29(10):906–915.

Elhindi KM, Al-Mana FA, El-Hendawy S, Al-Selwey WA, Elgorban AM. 2018. Arbuscular mycorrhizal fungi mitigates heavy metal toxicity adverse effects in sewage water contaminated soil on *Tagetes erecta* L. *Soil Science and Plant Nutrition* 64:662–668.

Etesami H. 2018. Bacterial mediated alleviation of heavy metal stress and decreased accumulation of metals in plant tissues: Mechanisms and future prospects. *Ecotoxicology and Environmental Safety* 147:175–191.

Faroon O, Ruiz P. 2016. Polychlorinated biphenyls: New evidence from the last decade. *Toxicology and Industrial Health* 32(11):1825–1847.

Fatima K, Afzal M, Imran A, Khan QM. 2015. Bacterial rhizosphere and endosphere populations associated with grasses and trees to be used for phytoremediation of crude oil contaminated soil. *Bulletin Environmental Contamination and Toxicology* 94:314–320

Fatima K, Imran A, Amin I, Khan QM, Afzal M. 2018. Successful phytoremediation of crude-oil contaminated soil at an oil exploration and production company by plants-bacterial synergism. *International Journal Of Phytoremediation* 20(7):675–681.

Fecih T, Baoune H. 2019. Arbuscular mycorrhizal fungi remediation potential of organic and inorganic compounds. In: Arora P (ed.) *Microbial Technology for the Welfare of Society*. *Microorganisms for Sustainability*, vol. 17. Singapore: Springer.

Fernández M, Niqui-Arroyo JL, Conde S, Ramos JL, Duque E. 2012. Enhanced tolerance to naphthalene and enhanced rhizoremediation performance for *Pseudomonas putida* KT2440 via the NAH7 catabolic plasmid. *Applied and Environmental Microbiology* 78(15):5104–5110.

Finkel OM, Castrillo G, Herrera Paredes S, Salas González I, Dangl JL. 2017. Understanding and exploiting plant beneficial microbes. *Current Opinion in Plant Biology* 38:155–163.

Foster RC. 1988. Microenvironments of soil microorganisms. *Biology and Fertility of Soils* 6:189–203.

Fouts DE, Tyler HL, DeBoy RT et al. 2008. Complete genome sequence of the N_2-Fixing broad host range endophyte *Klebsiella pneumoniae* 342 and virulence predictions verified in Mice. *PLoS Genetics* 4(7):e1000141.

Gadd GM. 2004. Microbial influence on metal mobility and application for bioremediation. *Geoderma* 122(2–4):109–119.

Gage DJ. 2004. Infection and invasion of roots by symbiotic, nitrogen-fixing rhizobia during nodulation of temperate legumes. *Microbiology and Molecular Biology Reviews* 68(2):280–300.

Gamalero E, Glick BR. 2012. Plant growth-promoting bacteria and metals phytoremediation. In: Anjum NA, Pereira ME, Ahmad I, Duarte AC, Umar S, Khan NA (eds). *Phytotechnologies: Remediation of Environmental Contaminants.* Boca Raton: CRC Press.

Ganesan V. 2008. Rhizoremediation of cadmium soil using a cadmium-resistant plant growth-promoting rhizopseudomonad. *Current Microbiology* 56(4):403–407.

Garbayetansley J. 1994. Review No. 76 Helper bacteria: A new dimension to the mycorrhizal symbiosis. *New Phytologist* 128(2):197–210.

Garbiscu C, Alkorta I. 2001. Phytoextraction: A cost effective plant based technology for the removal of metals from the environment. *Bioresource Technology* 77:229–236.

Garbisu C, Hernández-Allica J, Barrutia O, Alkorta I, Becerril JM. 2002. Phytoremediation: A technology using green plants to remove contaminants from polluted areas. *Reviews on Environmental Health* 17(3):173–188.

Garcia J, Kao-Kniffin J. 2018. Microbial group dynamics in plant rhizospheres and their implications on nutrient cycling. *Frontiers in Microbiology* 9:1516.

García-Sánchez M, Mercl F, Tlustoš P. 2018. A comparative study to evaluate natural attenuation, mycoaugmentation, phytoremediation, and microbial-assisted phytoremediation strategies for the bioremediation of an aged PAH-polluted soil. *Ecotoxicology and Environmental Safety* 147:165–174.

Gaur A, Adholeya A. 2004. Prospects of arbuscular mycorrhizal fungi in phytoremediation of heavy metal contaminated soils. *Current Science* 86(4):528–534.

Gerhardt KE, Gerwing PD, Huang X, Greenberg BM. 2015. Microbe-assisted phytoremediation of petroleum impacted soil: A scientifically proven green technology. In: Fingas M (ed.) *Handbook of Oil Spill Science and Technology.* John Wiley & Sons, Inc.

Germaine K, Keogh E, Garcia-Cabellos G et al. 2004. Colonisation of poplar trees by gfp expressing bacterial endophytes. *FEMS Microbiology Ecology* 48:109–118.

Germaine KJ, Keogh E, Ryan D, Dowling D. 2009. Bacterial endophyte-mediated naphthalene phytoprotection and phytoremediation. *FEMS Microbiology Letters* 296:226–234.

Germaine KJ, Liu X, Cabellos GG, Hogan JP, Ryan D, Dowling DN. 2006. Bacterial endophyte enhanced phytoremediation of the organochlorine herbicide 2, 4-dichlorophenoxyacetic acid. *FEMS Microbiology Ecology* 57:302–310.

Geurts R, Bisseling T. 2002. Rhizobium Nod factor perception and signalling. *Plant Cell* 14(Suppl):s239–s249.

Gkorezis P, Daghio M, Franzetti A, Van Hamme JD, Sillen W, Vangronsveld J. 2016. The interaction between plants and bacteria in the remediation of petroleum hydrocarbons: An environmental perspective. *Frontiers in Microbiology* 7:1836.

Glick BR. 1995. The enhancement of plant growth by free living bacteria. *Canadian Journal of Microbiology* 41:109–117.

Glick BR. 2003. Phytoremediation: Synergistic use of plants and bacteria to clean up the environment. *Biotechnology Advances* 21:383–393.

Glick BR. 2005. Modulation of plant ethylene levels by the bacterial enzyme ACC deaminase. *FEMS Microbiology Letters* 251(1):1–7.

Glick BR. 2010. Using soil bacteria to facilitate phytoremediation. *Biotechnology Advances* 28:367–374.

Glick BR. 2012. Plant growth-promoting bacteria: Mechanisms and applications. *Scientifica* 2012:15. Article ID 963401.

Glick BR. 2014. Bacteria with ACC deaminase can promote plant growth and help to feed the world. *Microbiology Research* 169:30–39.

Gontia-Mishra I, Sapre S, Sharma A, Tiwari S. 2016. Amelioration of drought tolerance in wheat by the interaction of plant growth-promoting rhizobacteria. *Plant Biology* 18(6):992–1000.

Gonzalez-Chavez C, D'Haen J, Vangronsveld J, Dodd J. 2002. Copper sorption and accumulation by the extraradical mycelium of different *Glomus* spp. (arbuscular mycorrhizal fungi) isolated from the same polluted soil. *Plant and Soil* 240:287–297.

Gonzalez-Chavez MC, Carrillo-Gonzalez R, Gutierrez-Castorena MC. 2009. Natural attenuation in a slag heap contaminated with cadmium: The role of plants and arbuscular mycorrhizal fungi. *Journal of Hazardous Materials* 161:1288–1298

González-Chávez MC, Carrillo-González R, Wright SF, Nichols KA. 2004. The role of glomalin, a protein produced by arbuscular mycorrhizal fungi, in sequestering potentially toxic elements. *Environmental Pollution* 130(3):317–323.

González-Chávez MDCA, Carrillo-González R, Cuellar-Sánchez A et al. 2019. Phytoremediation assisted by mycorrhizal fungi of a Mexican defunct lead-acid battery recycling site. *Science of the Total Environment* 650:3134–3144.

Gouda S, Kerry RG, Das G, Paramithiotis S, Shin HS, Patra JK. 2018. Revitalization of plant growth promoting rhizobacteria for sustainable development in agriculture. *Microbiological Research* 206:131–140.

Gray EJ, Smith DL. 2005. Intracellular and extracellular PGPR: Commonalities and distinctions in the plant–bacterium signaling processes. *Soil Biology and Biochemistry* 37:395–412.

Grobelak A, Napora A. 2015. The chemophytostabilisation process of heavy metal polluted soil. *PLOS ONE* 10(6):e0129538.

Gulati A, Sharma N, Vyas P et al. 2010. Organic acid production and plant growth promotion as a function of phosphate solubilization by *Acinetobacter* rhizosphaerae strain BIHB 723 isolated from the cold deserts of the trans-Himalayas. *Archives of Microbiology* 192:975–983.

Gullap MK, Dasci M, Erkovan HI, Koc A, Turan M. 2014. Plant growth-promoting rhizobacteria (pgpr) and phosphorus fertilizer-assisted phytoextraction of toxic heavy metals from contaminated soils. *Communications in Soil Science and Plant Analysis* 45(19):2593–2606.

Guo D, Fan Z, Lu S et al. 2019. Changes in rhizosphere bacterial communities during remediation of heavy metal-accumulating plants around the Xikuangshan mine in southern China. *Scientific Reports* 9:1947.

Guo JK, Ding YZ, Feng RW et al. 2015. *Burkholderia metalliresistens* sp. nov., a multiple metal-resistant and phosphate-solubilising species isolated from heavy metal-polluted soil in Southeast China. *Antonie Van Leeuwenhoek* 107(6):1591–1598.

Gupta P, Diwan B. 2017. Bacterial exopolysaccharide mediated heavy metal removal: A Review on biosynthesis, mechanism and remediation strategies. *Biotechnology Reports* 13:58–71.

Hadi F, Bano A. 2010. Effect of diazotrophs (*Rhizobium* and *Azatebactor*) on growth of maize (*Zea mays* L.) and accumulation of lead (Pb) in different plant parts. *Pakistan Journal of Botany* 42(6):4363–4370.

Hallmann J. 2001. Plant interactions with endophytic bacteria. In: Jeger MJ, Spence NJ (eds) *Biotic Interactions in Plant-Pathogen Associations*. Wallingford, UK: CABI, pp 87–119.

Hallmann J, Quadt-Hallmann A, Mahaffee WF, Kloepper JW. 1997. Bacterial endophytes in agricultural crops. *Canadian Journal Microbiology* 43:895–914.

Halvorson HO, Keynan A, Kornberg HL. 1990. Utilization of calcium phosphates for microbial growth at alkaline pH. *Soil Biology and Biochemistry* 22:887–890.

Hamdali H, Hafidi M, Virolle MJ, Ouhdouch Y. 2008. Rock phosphate-solubilizing Actinomycetes: Screening for plant growth-promoting activities. *World Journal of Microbiology and Biotechnology* 24:2565–2575.

Hardoim PR, van Overbeek LS, Elsas JD. 2008. Properties of bacterial endophytes and their proposed role in plant growth. *Trends in Microbiology* 16(10):463–471.

Hartmann A, Schmid M, van Tuinen D, Berg G. 2009. Plant-driven selection of microbes. *Plant and Soil* 321:235–257.

Hashim MA, Mukhopadhyay S, Sahu JN, Sengupta B. 2011. Remediation technologies for heavy metal contaminated groundwater. *Journal of Environmental Management* 92(10):2355–2388.

Hassan SE, Hijri M, St-Arnaud M. 2013. Effect of arbuscular mycorrhizal fungi on trace metal uptake by sunflower plants grown on cadmium contaminated soil. *Nature Biotechnology* 30(6):780–787.

Hassani MA, Durán P, Hacquard S. 2018. Microbial interactions within the plant holobiont. *Microbiome* 6:58.

Hata S, Kobae Y, Banba M. 2010. Interactions between plants and arbuscular mycorrhizal fungi. *International Review of Cell and Molecular Biology* 281:1–48.

He CQ, Tan GE, Liang X et al. 2010a. Effect of Zn-tolerant bacterial strains on growth and Zn accumulation in *Orychophragmus violaceus*. *Applied Soil Ecology* 44(1):1–5.

He H, Ye Z, Yang D, Yan J, Xiao L, Zhong T, Yuan M, Cai X, Fang Z, Jing Y. 2013. Characterization of endophytic *Rahnella* sp. JN6 from Polygonum pubescens and its potential in promoting growth and Cd, Pb, Zn uptake by *Brassica napus*. *Chemosphere* 90(6):1960–1965.

He LY, Chen ZJ, Ren GD, Zhang YF, Qian M, Sheng XF. 2009. Increased cadmium and lead uptake of a cadmium hyperaccumulator tomato by cadmium-resistant bacteria. *Ecotoxicology and Environmental Safety* 72(5):1343–1348.

He LY, Zhang YF, Ma HY et al. 2010b. Characterization of copper-resistant bacteria and assessment of bacterial communities in rhizosphere soils of copper-tolerant plants. *Applied Soil Ecology* 44:49–55.

Headley TR, Tanner CC. 2012. Constructed wetlands with floating emergent macrophytes: An innovative stormwater treatment technology. *Critical Reviews in Environmental Science and Technology* 42(21):2261–2310.

Hiltner L. 1904a. Uber neuere Erfahrungen und Probleme auf dem Gebiete der Bodenbakteriologie unter besonderden berucksichtigung und Brache. Arb. Dtsch. Landwirtsch. *Gesellschaft* 98:59–70.

Hiltner L. 1904b. Uber neuere Erfahrungen und Probleme auf dem Gebiete der Bodenbakteriologie unter besonderden berucksichtigung und Brache. Arb. Dtsch. Landwirtsch. *Gesellschaft* 98:59–78.

Ho YN, Mathew DC, Hsiao SC, Shih CH, Chien MF, Chiang HM, Huang CC. 2012. Selection and application of endophytic bacterium *Achromobacter xylosoxidans* strain F3B for improving phytoremediation of phenolic pollutants. *Journal of Hazardous Materials* 15:43–49.

Hodge A. 2014. Interactions between arbuscular mycorrhizal fungi and organic material substrates. *Advances in Applied Microbiology* 89:47–99.

Hodko D, Hyfte JV, Denvir A, Magnuson JW. 2000. Methods for enhancing phytoextraction of contaminants from porous media using electrokinetic phenomena. *US Patent No* 6:145–244.

Hossain MA, Piyatida P, da Silva JAT, Fujita M. 2012. Molecular mechanism of heavy metal toxicity and tolerance in plants: Central role of glutathione in detoxification of reactive oxygen species and methylglyoxal and in heavy metal chelation. *Journal of Botany* Article ID 872875. https://doi.org/10.1155/2012/872875

Hou J, Liu W, Wang B, Wang Q, Luo Y, Franks AE. 2015. PGPR enhanced phytoremediation of petroleum contaminated soil and rhizosphere microbial community response. *Chemosphere* 138:592–598.

Hrynkiewicz K, Złoch M, Kowalkowski T, Baum C, Buszewski B. 2018. Efficiency of microbially assisted phytoremediation of heavy-metal contaminated soils. *Environmental Reviews* 26(3):316–332.

Hubbell DH, Kidder G. 2009. Biological nitrogen fixation. *University of Florida IFAS Extension Publication SL* 16:1–4

Huertas E, Folch M, Salgot M, Gonzalvo I, Passarell C. 2006. Constructed wetlands effluent for streamflow augmentation in the Besòs River (Spain). *Desalination* 188(1–3):141–147.

Huo W, Zhuang C-H, Cao Y, Pu M, Yao H, Lou L-Q, Cai Q-S. 2012. Paclobutrazol and plant-growth promoting bacterial endophyte *Pantoea* sp. enhance copper tolerance of guinea grass (*Panicum maximum*) in hydroponic culture. *Acta Physiologiae Plantarum* 34:139–150.

Hussain Z, Arslan M, Malik MH, Mohsin M, Iqbal S, Afzal M. 2018. Treatment of the textile industry effluent in a pilot-scale vertical flow constructed wetland system augmented with bacterial endophytes. *Science of the Total Environment* 645:966–973.

Hussein A, Scholz M. 2017. Dye wastewater treatment by vertical-flow constructed wetlands. *Ecological Engineering* 101:28–38.

Hussein HS. 2008. Optimization of plant-bacteria complex for phytoremediation of contaminated soils. *International Journal of Botany* 4(4):437–443.

Hütsch BW, Augustin J, Merbach W. 2002. Plant rhizodeposition—an important source for carbon turnover in soils. *Journal of Plant Nutrition and Soil Science* 165(4).

Idris R, Trifonova R, Puschenreiter M, Wenzel WW, Sessitsch A. 2004. Bacterial communities associated with flowering plants of the ni hyperaccumulator *Thlaspi goesingense*. *Applied and Environmental Microbiology* 70(5):2667–2677.

Igbinosa EO, Odjadjare EE, Chigor VN, Igbinosa IH, Emoghene AO, Ekhaise FO, Igiehon NO, OG Idemudia. 2013. Toxicological Profile of Chlorophenols and Their Derivatives in the Environment: The Public Health Perspective. *The Scientific World Journal* Article ID 460215. http://dx.doi.org/10.1155/2013/460215

Ijaz A, Shabir G, Khan QM, Afzal M. 2015. Enhanced remediation of sewage effluent by endophyte-assisted floating treatment wetlands. *Ecological Engineering* 84:58–66.

Ike A, Sriprang R, Ono H, Murooka Y, Yamashita M. 2007. Bioremediation of cadmium contaminated soil using symbiosis between leguminous plant and recombinant rhizobia with the MTL4 and the PCS genes. *Chemosphere* 66:1670–1676.

Illmer P, Schinner F. 1995. Solubilization of inorganic calcium phosphates—Solubilization mechanisms. *Soil Biology and Biochemistry* 27(3):257–263.

Ionescu M, Beranova K, Dudkova V, Kochankova L, Demnerova K, Macek T, Mackova M. 2009. Isolation and characterization of different plant associated bacteria and their potential to degrade polychlorinated biphenyls. *International Biodeterioration & Biodegradation* 63:667–672.

Jacobsen C. 1997. Plant protection and rhizosphere colonization of barley by seed inoculated herbicide degrading *Burkholderia (Pseudomonas) cepacia* DBO1 (pRO101) in 2,4-D contaminated soil. *Plant and Soil* 189:139–144.

Jaishankar M, Tseten T, Anbalagan N, Mathew BB, Beeregowda KN. 2014. Toxicity, mechanism and health effects of some heavy metals. *Interdisciplinary Toxicology* 7(2):60–72.

Jambon I, Thijs S, Weyens N, Vangronsveld J. 2018. Harnessing plant-bacteria-fungi interactions to improve plant growth and degradation of organic pollutants. *Journal of Plant Microbe Interactions* 13(1):119–130.

James EK, Gyaneshwar P, Mathan N, Barraquio WL, Reddy PM, Iannetta PPM, Olivares FL, Ladha JK. 2002. Infection and colonization of rice seedlings by the plant growth-promoting bacterium *Herbaspirillum seropedicae* Z67. *Molecular Plant Microbe Interaction* 15:894–906.

Jan R, Khan MA, Asaf S, Lubna, Lee IJ, Kim KM. 2019. Metal resistant endophytic bacteria reduces cadmium, nickel toxicity, and enhances expression of metal stress related genes with improved growth of *Oryza Sativa*, via regulating its antioxidant machinery and endogenous hormones. *Plants* 8(10):363.

Järup L. 2003. Hazards of heavy metal contamination. *British Medical Bulletin* 68:167–182.

Jha P, Jha P. 2015. Plant-microbe partnerships for enhanced biodegrada- tion of polychlorinated piphenyls. In: Arora NK (ed) *Plant Microbes Symbiosis. Applied Facets.* India: Springer, pp. 95–110.

Jia X, Zhao Y, Liu T, Huang S, Chang Y. 2016. Elevated CO_2 increases glomalin-related soil protein (GRSP) in the rhizosphere of *Robinia pseudoacacia* L. seedlings in Pb- and Cd-contaminated soils. *Environmental Pollutition* 218:349–357.

Jian L, Bai X, Zhang H, Song X, Li Z. 2019. Promotion of growth and metal accumulation of alfalfa by coinoculation with *Sinorhizobium* and *Agrobacterium* under copper and zinc stress. *Peer Journal* 7:e6875.

Jiang CY, Sheng XF, Qian M, Wang QY. 2008. Isolation and characterization of a heavy metal-resistant *Burkholderia* sp. from heavy metal-contaminated paddy field soil and its potential in promoting plant growth and heavy metal accumulation in metal-polluted soil. *Chemosphere* 72(2):157–164.

Jing Y, He Z, Yang X. 2007. Role of soil rhizobacteria in phytoremediation of heavy metal contaminated soils. *Journal of Zhejiang Univ Science B* 8:192–207.

Jing YX, Yan JL, He HD, Yang DJ, Xiao L, Zhong T, Yuan M, Cai XD, Li SB. 2014. Characterization of bacteria in the rhizosphere soils of *Polygonum pubescens* and their potential in promoting growth and Cd, Pb, Zn uptake by *Brassica napus. Internaltional Journal of Phytoremediation* 16(4):321–333.

Johansson JF, Paul LR, Finlay RD. 2004. Microbial interactions in the mycorrhizosphere and their significance for sustainable agriculture. *FEMS Microbiol Ecology* 48(1):1–13.

Joner E, Briones R, Leyval C. 2000. Metal-binding capacity of arbuscular mycorrhizal mycelium. *Plant and Soil* 226:227–234.

Joner EJ, Leyval C. 1997. Uptake of ^{109}Cd by roots and hyphae of a *Glomus mosseae/Trifolium subterraneum* mycorrhiza from soil amended with high and low concentrations of cadmium. *New Phytologist* 135(2):353–360.

Joner EJ, Leyval C. 2001. Influence of arbuscular mycorrhiza on clover and ryegrass grown together in a soil spiked with polycyclic aromatic hydrocarbons. *Mycorrhiza* 10:155–159.

Joner EJ, Leyval C. 2003. Rhizosphere gradients of polycyclic aromatic hydrocarbon (PAH) dissipation in two industrial soils and the impact of arbuscular mycorrhiza. *Environmental Science and Technology* 37(11):2371–2375.

Jones D, Smith BFL, Wilson MJ, Goodman BA.1991. Phosphate solubilizing fungi in a Scottish upland soil. *Mycological Research* 95:1090–1903.

Joshi PM, Juwarkar AA. 2009. In vivo studies to elucidate the role of extracellular polymeric substances from Azotobacter in immobilization of heavy metals. *Environmental Science & Technology* 43(15):5884–5889.

Juederie OBO, Babalola OO. 2017. Microbial and plant-assisted bioremediation of heavy metal polluted environments: A Review. *International Journal of Environmental Research and Public Health* 14:1504.

Juhanson J, Truu J, Heinaru E, Heinaru A. 2007. Temporal dynamics of microbial community in soil during phytoremediation field experiment. *Journal of Environmental Engineering and Landscape Management* 15(4):213–220.

Juwarkar AA, Nair A, Dubey KV, Singh SK, Devotta S. 2007. Biosurfactant technology for remediation of cadmium and lead contaminated soils. *Chemosphere* 68(10):1996–2002.

Juwarkar AA, Singh SK. 2010. Microbe-assisted phytoremediation approach for ecological restoration of zinc mine spoil dump. *International Journal of Environment and Pollution* 43(1/2/3):236–250.

Kabeer R, Varghese R, Kannan VM, Thomas JR, Poulose SV. 2014. Rhizosphere bacterial diversity and heavy metal accumulation in *Nymphaea pubescens* in aid of phytoremediation potential. *Journal of Bioscience and Biotechnology* 3(1):89–95.

Kabra AN, Khandare RV, Govindwar SP. 2013. Development of a bioreactor for remediation of textile effluent and dye mixture: A plant-bacterial synergistic strategy. *Water Research* 47:1035–1048.

Kadlec RH, Knight RL. 1996. *Treatment Wetlands*. New York: Lewis Publishers, CRC.

Kadlec RH, Knight RL, Vyamazal J, Brix H, Cooper P, Haberl, R. 2000. *Constructed Wetlands for Pollution Control-Processes, Performance Design and Operation*. IWA Scientific and Technical Report No. 8. IWA, Publishing, London.

Kadlec RH, Wallace S. 2008. *Treatment Wetlands*. 2nd Edition, CRC Press.

Kakimoto T. 2003. Perception and signal transduction of cytokinins. *Annual Review of Plant Biology* 54:605–627.

Kalayu G. 2019. Phosphate solubilizing microorganisms: Promising approach as biofertilizers. *International Journal of Agronomy* 2019:7. Article ID 4917256.

Kalia VC, Chauhan A, Bhattacharyya G. 2003. Genomic databases yield novel bioplastic producers. *Nature Biotechnology* 21:845–846.

Kang J, Amoozegar A, Hesterberg D, Osmond DL. 2011. Phosphorus leaching in a sandy soil as affected by organic and inorganic fertilizer sources. *Geoderma* 161(3–4):194–201.

Kang JW, Khan Z, Doty SL. 2012. Biodegradation of trichloroethylene (TCE) by an endophyte of hybrid popular. *Applied and Environmental Microbiology* 78:3504–3507.

Karthikeyan R, Kulakow PA. 2003. Soil plant microbe interactions in phytoremediation. *Advances in Biochemical Engineering and Biotechnology* 78:51–74.

Katrin S, Headley JV, Peru KM, Haidar N, Gurprasard NP. 2005. Residues of explosives in groundwater leached from soils at a military site in eastern Germany. *Communications in Soil Science and Plant Analysis 36, 2005—Issue 1–3: The 8th International Symposium on Soil and Plant Analysis: Part One*, 133–153.

Khan AA, Jilani G, Akhter MS, Naqvi SMS, Rasheed M. 2016. Phosphorous solubilizing Bacteria; occurrence, Mechanisms and their role in crop production. *Journal of Agricultural and Biological Science* 1:48–58.

Khan AG. 2005. Role of soil microbes in the rhizospheres of plants growing on trace metal contaminated soils in phytoremediation. *Journal of Trace Elements in Medicine and Biology* 18(4):355–364.

Khan AG. 2006. Mycorrhizoremediation—an enhanced form of phytoremediation. *Journal of Zhejiang Univ Science B* 7(7):503–514.

Khan AL, Waqas M, Kang S-M et al. 2014a. Bacterial endophyte *Sphingomonas* sp. LK11 produces gibberellins and IAA and promotes tomato plant growth. *Journal of Microbiology* 52:689–695.

Khan AR, Ullah I, Khan AL, Park G-S, Waqas M, Hong S-J, Jung BK, Kwak Y, Lee I-J, Shin J-H. Improvement in phytoremediation potential of Solanum nigrum under cadmium contamination through endophytic-assisted Serratia sp. RSC-14 inoculation. *Environmental Science and Pollution Research International* 22(18):14032–14042.

Khan MS, Zaidi A, Wani PA, Oves M. 2009. Role of plant growth promoting rhizobacteria in the remediation of metal contaminated soils: A review. In: Lichtfouse E (ed) *Organic Farming, Pest Control and Remediation of Soil Pollutants. Sustainable Agriculture Reviews*, vol 1. Dordrecht: Springer.

Khan MU, Sessitsch A, Harris M et al. 2014b. Cr-resistant rhizo- and endophytic bacteria associated with *Prosopis juliflora* and their potential as phytoremediation enhancing agents in metal-degraded soils. *Frontiers in Plant Science* 1(755):1–10.

Khan S, Afzal M, Iqbal S, Khan QM. 2013. Plant–bacteria partnerships for the remediation of hydrocarbon contaminated soils. *Chemosphere* 90:1317–1332.

Khan Z, Roman D, Kintz T, delas Alas M, Yap R, Doty S. 2014a. Degradation, phytoprotection and phytoremediation of phenanthrene by endophyte *Pseudomonas putida* PD1. *Environmental Science and Technology* 48(20):12221–12228.

Khandare RV, Kabra AN, Kadam AA, Govindwar SP. 2013. Treatment of dye containing wastewaters by a developed lab scale phytoreactor and enhancement of its efficacy by bacterial augmentation. *International Biodeterioration & Biodegradation* 78:89–97.

Khandare RV, Rane NR, Waghmode TR, Govindwar SP. 2012. Bacterial assisted phytoremediation for enhanced degradation of highly sulfonated diazo reactive dye. *Environmental Science and Pollution Research* 19:1709–1718.

Kim KR, Owens G, Naidu R. 2010. Effect of root-induced chemical changes on dynamics and plant uptake of heavy metals in rhizosphere soils. *Pedosphere* 20(4):494–504.

Kloepper JW, Ryu C-M, Zhang SA. 2004. Induced systemic resistance and promotion of plant growth by *Bacillus* spp. *Phytopathology* 94:1259–1266.

Kloepper JW, Schroth MN. 1978. Plant growth-promoting rhizobacteria on radishes. *Proceedings of the 4th International Conference on Plant Pathogenic Bacteria, vol. 2. Station de Pathologie Végétale et de Phytobactériologie, INRA*, Angers, France, 879–882.

Knoth JL, Kim SH, Ettl GJ, Doty SL. 2013. Effects of cross host species inoculation of nitrogen-fixing endophytes on growth and leaf physiology of maize. *GCB Bioenergy* 5:408–418.

Kong Z, Glick BR. 2017. The role of plant growth-promoting bacteria in metal phytoremediation. *Advances in Microbial Physiology* 71:97–132.

Kovtunovych G, Lar O, Kamalova S, Kordyum V, Kleiner D, Kozyrovska N. 1999. Correlation between pectate lyase activity and ability of diazotrophic *Klebsiella oxytoca* VN 13 to penetrate into plant tissues. *Plant and Soil* 215:1–6.

Kubiak JJ, Khankhane PJ, Kleingeld PJ, Lima AT. 2012. An attempt to electrically enhance phytoremediation of arsenic contaminated water. *Chemosphere* 87:259–264.

Kuffner M, De Maria S, Puschenreiter M et al. 2010. Culturable bacteria from Zn- and Cd-accumulating *Salix caprea* with differential effects on plant growth and heavy metal availability. *Journal of Applied Microbiology* 108:1471–1484.

Kuffner M, Puschenreiter M, Wieshammer G, Gorfer M, Sessitsch A. 2008. Rhizosphere bacteria affect growth and metal uptake of heavy metal accumulating willows. *Plant and Soil* 304:35–44.

Kuiper I, Lagendijk EL, Bloemberg GV, Lugtenberg BJJ. 2004. Rhizoremediation: A Beneficial Plant-Microbe Interaction. *Molecular Plant-Microbe Interactions* 17(1):6–15.

Kumar KV, Singh N, Behl HM, Srivastava S. 2008. Influence of plant growth promoting bacteria and its mutant on heavy metal toxicity in *Brassica juncea* grown in fly ash amended soil. *Chemosphere* 72(4):678–683.

Kumar KV, Srivastava S, Singh N, Behl HM. 2009. Role of metal resistant plant growth promoting bacteria in ameliorating fly ash to the growth of *Brassica juncea*. *Journal of Hazardous Materials* 170(1):51–57.

Kumar V, Chandra R. 2018a. Bacterial assisted phytoremediation of industrial waste pollutants and eco-restoration. In: Chandra R, Dubey NK, Kumar V (eds) *Phytoremediation of Environmental Pollutants*. Boca Raton: CRC Press.

Kumar V, Chandra R. 2018b. Characterisation of manganese peroxidase and laccase producing bacteria capable for degradation of sucrose glutamic acid-maillard products at different nutritional and environmental conditions. *World Journal of Microbiology and Biotechnology* 34:32.

Kumar V, Chandra R. 2020a. Bioremediation of melanoidins containing distillery waste for environmental safety. In: Saxena G, Bharagava RN (eds) *Bioremediation of Industrial Waste for Environmental Safety. Vol II- Microbes and Methods for Industrial Waste Management*. Singapore: Springer.

Kumar V, Chandra R. 2020b. Bacterial-assisted phytoextraction mechanism of heavy metals by native hyperaccumulator plants from distillery waste-contaminated site for eco-restoration. In: Chandra R, Sobti R (eds) *Microbes for Sustainable Development and Bioremediation*. Boca Raton: CRC Press.

Kumar V, Chandra R, Thakur IS, Saxena G, Shah MP. 2020. Recent advances in physicochemical and biological treatment approaches for distillery wastewater. In: Shah M, Banerjee A (eds) *Combined Application of Physico-Chemical & Microbiological Processes for Industrial Effluent Treatment Plant.* https://doi.org/10.1007/978-981-15-0497-6_6

Kumar V, Shahi SK, Singh S. 2018. Bioremediation: An eco-sustainable approach for restoration of contaminated sites. In: Singh J, Sharma D, Kumar G, Sharma NR (eds) *Microbial Bioprospecting for Sustainable Development.* Singapore: Springer.

Kumar V, Sharma DC. 2019. Distillery effluent: Pollution profile, eco-friendly treatment strategies, challenges, and future prospects. In: Arora PK (eds) *Microbial Metabolism of Xenobiotic Compounds.* Springer Nature. https://doi.org/10.1007/978-981-13-7462-3_17.

Kurth F, Zeitler K, Feldhahn L, Neu TR, Weber T, Krištůfek V, Wubet T, Herrmann S, Buscot F, Tarkka MT. 2013. Detection and quantification of a mycorrhization helper bacterium and a mycorrhizal fungus in plant-soil microcosms at different levels of complexity. *BMC Microbiology* 13:205.

Labbé JL, Weston DJ, Dunkirk N, Pelletier DA, Tuskan GA. 2014. Newly identified helper bacteria stimulate ectomycorrhizal formation in Populus. *Frontiers in Plant Scieence* 5:579.

Lambolez L, Vasseur P, Ferard JF, Gisbert T. 1994. The environmental risks of industrial waste disposal: An experimental approach including acute and chronic toxicity studies. *Ecotoxicology and Environmental Safety* 28(3):317–328.

Lampel JS, Canter GL, Dimock MB, Kelly JL, Anderson JJ, Uratani BB, Foulke JS Jr, Turner JT. 1994. Integrative cloning, expression, and stability of the cryIA. gene from Bacillus thuringiensis subsp. kurstaki in a recombinant strain of *Clavibacter xyli* subsp. cynodontis. *Applied and Environmental Microbiology* 60:501–508.

Lareen A, Burton F, Schäfer P. 2016. Plant root-microbe communication in shaping root microbiomes. *Plant and Molecular Biology* 90:575–587.

Leandro dos Santos M, Soares CRFS, Comin JJ, Lovato PE. 2017. The phytoprotective effects of arbuscular mycorrhizal fungi on *Enterolobium contorstisiliquum* (Vell.) Morong in soil containing coal-mine tailings. *International Journal of Phytoremediation* 19(12):1100–1108.

Lee W, Wood TK, Chen W. 2006. Engineering TCE-degrading rhizobacteria for heavy metal accumulation and enhanced TCE degradation. *Biotechnology and Bioengineering* 95(3):399–403

Leigh MB, Prouzová P, Macková M, Macek T, Nagle DP, Fletcher JS. 2006. Polychlorinated biphenyl (PCB)-degrading bacteria associated with trees in a PCB-contaminated site. *Applied and Environmental Microbiology* 72(4):2331–2342.

Lemoigne M. 1926. Produits de dehydration et de polymerisation de l acide ßoxobutyrique. *Bull Soc Chim Biol* 8:770–782.

Lenoir I, LounesHadj Sahraoui A, Fontaine J. 2016. Arbuscular mycorrhizal fungal-assisted phytoremediation of soil contaminated with persistent organic pollutants: A review. *European Journal of Soil Science* 67(5):624–640.

Leung HM, Leung AOW, Ye ZH, Cheung KC, Yung KKL. 2013. Mixed arbuscular mycorrhizal (AM) fungal application to improve growth and arsenic accumulation of *Pteris vittata* (As hyperaccumulator) grown in As-contaminated soil. *Chemosphere* 92(10):1367–1374.

Leyval C, Joner EJ, del Val C, Haselwandter K. 2002. Potential of arbuscular mycorrhizal fungi for bioremediation. In: Gianinazzi S, Schüepp H, Barea JM, Haselwandter K. (eds) *Mycorrhizal Technology in Agriculture.* Basel: Birkhäuser.

Li K, Ramakrishna W. 2011. Effect of multiple metal resistant bacteria from contaminated lake sediments on metal accumulation and plant growth. *Journal of Hazardous Materials* 189(1–2):531–539.

Li Q, Ling W, Gao Y, Li F, Xiong W. 2006. Arbuscular mycorrhizal bioremediation and its mechanisms of organic pollutants-contaminated soils. *Ying Yong Sheng Tai Xue Bao* 17(11):2217–2221.

Li WC, Ye ZH, Wong MH. 2007. Effects of bacteria on enhanced metal uptake of the Cd/Zn-hyperaccumulating plant, *Sedum alfredii*. *Journal of Experimental Botany* 58(15–16):4173–4182.

Limmer MA, Burken JG. 2016. Phytovolatilization of organic contaminants. *Environmental Science and Technology* 50:6632–6643.

Lintern M, Anand R, Ryan C, Paterson D. 2013. Natural gold particles in *Eucalyptus* leaves and their relevance to exploration for buried gold deposits. *Nature Communications* 4:2614. https://doi.org/10.1038/ncomms3614

Lintern M, Anand R, Ryan C, Paterson D. 2018. Natural gold particles in *Eucalyptus* leaves and their relevance to exploration for buried gold deposits. *Nature Communications* 4:2614.

Liu L, Li W, Song W, Guo M. 2018. Remediation techniques for heavy metal-contaminated soils: Principles and applicability. *Science of the Total Environment* 633:206–219.

Liu W, Sun J, Ding L, Luo Y, Chen M, Tang C. 2013. Rhizobacteria (*Pseudomonas* sp. SB) assist phytoremediation of oily-sludge-contaminated soil by tall fescue (*Testuca arundinacea* L.). *Plant and Soil* 371:533–542.

Liu W, Wang Q, Wang B, Hou J, Luo Y, Tang C, Franks AE. 2015. Plant growth-promoting rhizobacteria enhance the growth and Cd uptke of *Sedum plumbizincicola* in a Cd-contaminated soil. *Journal of Soils Sediments* 15:1191–1199.

Lodewyckx C, Taghavi S, Mergeay M, Vangronsveld J, Clijsters H, van der Lelie D. 2002. The effect of recombinant heavy metal-resistant endophytic bacteria on heavy metal uptake by their host plant. *International Journal of Phytoremediation* 3(2):173–187.

López-Bucio J, Campos-Cuevas JC, Hernández-Calderón E, Velásquez-Becerra C, Farías-Rodríguez R, Macías-Rodríguez LI, Valencia-Cantero E. 2007. *Bacillus megaterium* rhizobacteria promote growth and alter root-system architecture through an auxin- and ethylene-independent signaling mechanism in *Arabidopsis thaliana*. *Molecular Plant-Microbe Interactions* 20(2):207–217.

Lugtenberg B, Kamilova F. 2009. Plant-growth-promoting rhizobacteria. *Annual Reviews in Microbiology* 63:541–556.

Luo Q, Sun L, Hu X, Zhou R. 2014. The variation of root exudates from the hyperaccumulator *Sedum alfredii* under cadmium stress: Metabonomics analysis. *PLOS ONE* 9(12):e115581.

Luo S, Xu T, Chen L et al. 2012. Endophyte-assisted promotion of biomass production and metal-uptake of energy crop sweet sorghum by plant-growth-promoting endophyte *Bacillus* sp. SLS18. *Appllied Microbiology and Biotechnology* 93(4):1745–1753.

Luo SL, Chen L, Chen JI et al. 2011. Analysis and characterization of cultivable heavy metal-resistant bacterial endophytes isolated from Cd hyperaccumulator *Solanum nigrum* L. and their potential use for phytoremediation. *Chemosphere* 85:1130–1138.

Ma J, Bei Q, Wang X, Lan P et al. 2019. Impacts of Mo application on biological nitrogen fixation and diazotrophic communities in a flooded rice-soil system. *Science of the Total Environment* 649:686–694.

Ma Y, Oliveira RS, Freitas H, Zhang C. 2016a. Biochemical and molecular mechanisms of plant-microbe-metal interactions: Relevance for phytoremediation. *Frontiers in Plant Science* 7:918.

Ma Y, Oliveira RS, Nai F, Rajkumar M, Luo Y, Rocha I, Freitas H. 2015. The hyperaccumulator *Sedum plumbizincicola* harbors metal-resistant endophytic bacteria that improve its phytoextraction capacity in multi-metal contaminated soil. *Journal of Environmental Management* 156:62–69.

Ma Y, Rajkumar M, Freitas H. 2009a. Improvement of plant growth and nickel uptake by nickel resistant-plant-growth promoting bacteria. *Journal of Hazardous Materials* 166(2–3):1154–1161.

Ma Y, Rajkumar M, Freitas H. 2009b. Inoculation of plant growth promoting bacterium *Achromobacter xylosoxidans* strain Ax10 for the improvement of copper phytoextraction by *Brassica juncea*. *Journal of Environmental Management* 90:831–837.

Ma Y, Rajkumar M, Luo Y, Freitas H. 2013. Phytoextraction of heavy metal polluted soils using *Sedum plumbizincicola* inoculated with metal mobilizing *Phyllobacterium myrsinacearum* RC6b. *Chemosphere* 93(7):1386–1392.

Ma Y, Rajkumar M, Zhang C, Freitas H. 2016b. Beneficial role of bacterial endophytes in heavy metal phytoremediation. *Journal of Environmental Management* 174:14–25.

Madhaiyan M, Poonguzhali S, Sa T. 2007. Metal tolerating methylotrophic bacteria reduces nickel and cadmium toxicity and promotes plant growth of tomato (*Lycopersicon esculentum* L.). *Chemosphere* 69(2):220–228

Mahdi SS, Hassan GI, Hussain A, Rasool F. 2011. Phosphorus availability issue-its fixation and role of phosphate solubilizing bacteria in phosphate solubilization. *Research Journal of Agricultural Science* 2:174–179.

Majumder A, Bhattacharyya K, Bhattacharyya S, Kole SC. 2013. Arsenic-tolerant, arsenite-oxidising bacterial strains in the contaminated soils of West Bengal, India. *Science of The Total Environment* 463–464:1006–1014.

Malik A. 2004. Metal bioremediation through growing cells. *Environmental International* 30(2):261–278.

Mannisto MK, Tiirola MA, Puhakka JA. 2001. Degradation of 2, 3,4,6-tetraclorophenol at low temperature and low dioxygen concentrations by phylogenetically different groundwater and bioreactor bacteria. *Biodegradation* 12:291–301.

Manzetti S. 2013. Polycyclic aromatic hydrocarbons in the environment: Environmental Fate and Transformation. *Polycyclic Aromatic Compounds* 33(4):311–330.

Manzoor M, Gul I, Kallerhoff J, Arshad M. 2019. Fungi-assisted phytoextraction of lead: Tolerance, plant growth–promoting activities and phytoavailability. *Environmental Science and Pollution Research* 26:23788–23797.

Mao X, Han FX, Shao X, Guo K, McComb J, Arslan Z, Zhang Z. 2016. Electro-kinetic remediation coupled with phytoremediation to remove lead, arsenic and cesium from contaminated paddy soil. *Ecotoxicology and Environmental Safety* 125:16–24.

Maqbool F, Wang Z, Xu Y, Zhao J, Gao D, Zhao YG, Bhatti ZA, Xing B. 2012. Rhizodegradation of petroleum hydrocarbons by *Sesbania cannabina* in bioaugmented soil with free and immobilized consortium. *Journal of Hazardous Materials* 237–238:262–269.

Marathe RJ, Phatake YB, Shaikh AC, Shinde BP, Gajbhiye MH. 2017. Effect of IAA produced by *Pseudomonas aeruginosa* 6a (bc4) on seed germination and plant growth of *Glycin max*. *Journal of Experimental Biology and Agricultural Sciences* 5(3). http://dx.doi.org/10.18006/2017.5(3).351.358

Marchut-Mikolajczyk O, Drożdżyński P, Pietrzyk D et al. 2018. Biosurfactant production and hydrocarbon degradation activity of endophytic bacteria isolated from *Chelidonium majus* L. *Microbial Cell Factories* 17:171.

Marques APGC, Rangel AOSS, Castro PML. 2009. Remediation of heavy metal contaminated soils: Phytoremediation as a potentially promising clean-up technology. *Critical Reviews in Environmental Science and Technology* 39(8).

Martinetti G, Loper JE. 1992. Mutational analysis of genes determining antagonism of Alcaligenes sp. strain MFA1 against the phytopathogenic fungus *Fusarium oxysporum*. *Canadian Journal of Microbiology* 38(3):241–247.

Matilla MA, Espinosa-Urgel M, Rodríguez-Herva JJ, Ramos JL, Ramos-González MI. 2007. Genomic analysis reveals the major driving forces of bacterial life in the rhizosphere. *Genome Biology* 8:R179.

McGrath SP, Zhao FJ. 2003. Phytoextraction of metals and metalloids from contaminated soils. *Current Opinion in Biotechnology* 14:277–282.

McGuinness M, Ivory C, Gilmartin N, Dowling DN. 2006. Investigation of substrate specificity of wildtype and mutant BphK[LB400] (a glutathione S-transferase) from Burkholderia LB400. *International Biodeterioration & Biodegradation* 58(3–4):203–208.

McGuinness MC, Mazurkiewicz V, Brennan E. 2007. Dechlorination of pesticides by a specific bacterial glutathione s-transferase, BphK[LB400]: Potential for bioremediation. *Engineering and Life Science* 7(6):611–615.

McNear Jr DH. 2013. The Rhizosphere—roots, soil and everything in between. *Nature Education Knowledge* 4(3):1.

Mehmannavaz R, Prasher SO, Ahmad D. 2002. Rhizospheric effects of alfalfa on biotransformation of polychlorinated biphenyls in a contaminated soil augmented with *Sinorhizobium meliloti*. *Process Biochemistry* 37(9):955–963.

Mehta P, Walia A, Kulshrestha S, Chauhan A, Shirkot CK. 2004.Efficiency of plant growth-promoting P-solubilizing *Bacillus circulans* CB7 for enhancement of tomato growth under net house conditions. *Journal of Basic Microbiology* 53:1–12.

Meier S. 2011. Contribution of metallophyte/arbuscular mycorrhizal symbiosis to the phytoremediation processes of copper contaminated soils. (Unpublished doctoral thesis). Universidad de la Frontera, Temuco, Chile.

Meier S, Borie F, Bolan N, Cornejo P. 2012. Phytoremediation of metal-polluted soils by arbuscular mycorrhizal fungi. *Critical Reviews in Environmental Science and Technology* 42(7).

Mendes R, Garbeva P, Raaijmakers JM. 2013. The rhizosphere microbiome: Significance of plant beneficial, plant pathogenic, and human pathogenic microorganisms. *FEMS Microbiology Reviews* 37:634–663.

Mercado-Blanco J, Abrantes I, Barra Caracciolo A et al. 2018. Belowground microbiota and the health of tree Crops. *Frontiers in Microbiology* 9:1006.

Mercado-Blanco J, Prieto P. 2012. Bacterial endophytes and root hairs. *Plant and Soil* 36:301–306.

Mesa J, Mateos-Naranjo E, Caviedes MA, Redondo-Gómez S, Pajuelo E, Rodríguez-Llorente ID. 2015. Endophytic cultivable bacteria of the metal bioaccumulator spartina maritima improve plant growth but not metal uptake in polluted marshes soils. *Frontiers in Microbiology* 6:450.

Mesa V, Navazas A, González-Gil R. 2017. Use of endophytic and rhizosphere bacteria to improve phytoremediation of arsenic-contaminated industrial soils by autochthonous *Betula celtiberica*. *Applied and Environmental Microbiology* 83:03411–03416.

Mhlongo MI, Piater LA, Madala NE, Labuschagne N, Dubery IA. 2018. The chemistry of plant-microbe interactions in the rhizosphere and the potential for metabolomics to reveal signaling related to defense priming and induced systemic resistance. *Frontiers in Plant Science* 9:112.

Miller CM, Miller RV, Garton-Kenny D, Redgrave B, Sears J, Condron MM, Teplow DB, Strobel GA. 1998. Ecomycins, unique antimycotics from *Pseudomonas viridiflava*. *Journal of Applied Microbiology* 84:937–944.

Miransari M. 2011. Hyperaccumulators, arbuscular mycorrhizal fungi and stress of heavy metals. *Biotechnology Advances* 29(6):645–653.

Mitter B, Petric A, Shin MW, Chain PS, Hauberg-Lotte L, Reinhold-Hurek B, Nowak J, Sessitsch A. 2013. Comparative genome analysis of *Burkholderia phytofirmans* PsJN reveals a wide spectrum of endophytic lifestyles based on interaction strategies with host plants. *Frontiers in Plant Science* 4:1–15.

Mitter EK, Kataoka R, de Freitas JR, Germida JJ. 2019. Potential use of endophytic root bacteria and host plants to degrade hydrocarbons. *International Journal of Phytoremediation* 21(9):928–938.

Montiel-Rozas MM, Madejón E, Madejón P. 2016. Effect of heavy metals and organic matter on root exudates (low molecular weight organic acids) of herbaceous species: An assessment in sand and soil conditions under different levels of contamination. *Environmental Pollution* 216:273–281.

Moore FP, Barac T, Borremans B et al. 2006. Endophytic bacterial diversity in poplar trees growing on a BTEX-contaminated site: The characterization of isolates with potential to enhance phytoremediation. *Systematic and Applied Microbiology* 29:539–556.

Moreira H, Marques AP, Franco AR, Rangel AO, Castro PM. 2014. Phytomanagement of Cd-contaminated soils using maize (*Zea mays* L.) assisted by plant growth-promoting rhizobacteria. *Environmental Science and Pollution Research International* 21(16):9742–9753.

Mosa KA, Saadoun I, Kumar K, Helmy M, Dhankher OP. 2016. Potential biotechnological strategies for the cleanup of heavy metals and metalloids. *Frontiers in Plant Science* 7:303.

Mukherjee G, Saha C, Naskar N et al. 2018. An Endophytic bacterial consortium modulates multiple strategies to improve arsenic phytoremediation efficacy in Solanum nigrum. *Scientific Reports* 8:6979.

Mumtaz MZ, Ahmad M, Jamil M, Hussain T. 2017. Zinc solubilizing *Bacillus* spp. potential candidates for biofortification in maize. *Microbiology Research* 202:51–60.

Muratova AY, Turkovskaya OV, Antonyuk LP, Makarov OE, Pozdnyakova LI, Ignatov VV. 2005. Oil-oxidizing potential of associative rhizobacteria of the genus *Azospirillum*. *Microbiology* 74:210–215.

Naghipour D, Ashrafi SD, Gholamzadeh M, Taghavi K, Naimi-Joubani M. 2018. Phytoremediation of heavy metals (Ni, Cd, Pb) by Azolla filiculoides from aqueous solution: A dataset. *Data Brief* 21:1409–1414.

Nanekar S, Dhote M, Kashyap S, Singh SK, Juwarkar AA. 2015. Microbe assisted phytoremediation of oil sludge and role of amendments: A mesocosm study. *International Journal of Environmental Science and Technology* 12:193–202.

Narasimhan K, Basheer C, Bajic VB, Swarup S. 2003. Enhancement of plant-microbe interactions using a rhizosphere metabolomics-driven approach and its application in the removal of polychlorinated biphenyls. *Plant Physiology* 132:146–153.

Nedelkoska TV, Doran PM. 2000. Characteristics of heavy metal uptake by plant species with potential for phytoremediation and phytomining. *Minerals Engineering* 13(5):549–561.

Newman LA, Reynol CM. 2005. Bacteria and phytoremediation: New uses for endophytic bacteria in plants. *Trends in Biotechnology* 23:6–8.

Newman LA, Reynolds CM. 2004. Phytodegradation of organic compounds. *Current Opinion in Biotechnology* 15(3):225–230.

Nicoară A, Neagoe A, Stancu P, De Giudici G, Langella F, Sprocati A, Iordache V, Kothe E. 2014. Coupled pot and lysimeter experiments assessing plant performance in microbially assisted phytoremediation. *Environmental Science Pollution Research* 21:6905–6920.

Nie M, Wang Y, Yu J et al. 2011. Understanding plant-microbe interactions for phytoremediation of petroleum-polluted soil. *PLOS ONE* 6(3):e17961.

Norman JS, Friesen ML. 2017. Complex N acquisition by soil diazotrophs: How the ability to release exoenzymes affects N fixation by terrestrial free-living diazotrophs. *The ISME Journal* 11:315–326.

Ojuederie OB, Babalola OO. 2017a. Microbial and plant-assisted bioremediation of heavy metal polluted environments: A review. *International Journal of Environmental Research and Public Health* 14(12). doi: 10.3390/ijerph14121504

Ojuederie OB, Babalola OO. 2017b. Microbial and Plant-Assisted Bioremediation of Heavy Metal Polluted Environments: A Review. *International Journal of Environmental Research and Public Health* 14(12):1504.

Olanrewaju OS, Ayangbenro AS, Glick BR, Babalola OO. 2019. Plant health: Feedback effect of root exudates-rhizobiome interactions. *Applied Microbiology and Biotechnology* 103:1155–1166.

Olanrewaju OS, Glick BR, Babalola OO. 2017. Mechanisms of action of plant growth promoting bacteria. *World Journal of Microbiology and Biotechnology* 33(11):197.

Oves M, Khan MS, Zaidi A. 2013. Chromium reducing and plant growth promoting novel strain *Pseudomonas aeruginosa* OSG41 enhance chickpea growth in chromium amended soils. *European Journal of Soil Biology* 56:72–83.

Pacheco GJ, Ciapina EMP, de Barros Gomes E, Junior NP. 2010.Biosurfactant production by Rhodococcus erythropolis and its application to oil removal. *Brazilian Journal of Microbiology* 41(3):685–693.

Pacwa-Płociniczak M, Płaza GA, Piotrowska-Seget Z, Cameotra SS. 2011. Environmental applications of biosurfactants: Recent advances. *International Journal of Molecular Sciences* 12(1):633–654.

Padmavathiamma PK, Li LY. 2007. Phytoremediation technology: Hyper-accumulation metals in plants. *Water Air Soil Pollution* 184:105.

Pal A, Paul AK. 2008. Microbial extracellular polymeric substances: Central elements in heavy metal bioremediation. *Indian Journal of Microbiology* 48(1):49–64.

Pal A, Wauters G, Paul AK. 2007. Nickel tolerance and accumulation by bacteria from rhizosphere of nickel hyperaccumulators in serpentine soil ecosystem of Andaman, India. *Plant and Soil* 293:37–48.

Pandey S, Ghosh PK, Ghosh S, De TK, Maiti TK. 2013. Role of heavy metal resistant *Ochrobactrum* sp. and *Bacillus* spp. strains in bioremediation of a rice cultivar and their PGPR like activities. *Journal of Microbiology* 51:11–17.

Panhwar QA, Jusop S, Naher UA, Othman R, Razi MI. 2013. Application of potential phosphate-solubilizing bacteria and organic acids on phosphate solubilization from phosphate rock in aerobic rice.*The ScientificWorld Journal* 2013:10. Article ID 272409.

Passatore L, Rossetti S, Juwarkar AA, Massacci A. 2014. Phytoremediation and bioremediation of polychlorinated biphenyls (PCBs): State of knowledge and research perspectives. *Journal of Hazardous Materials* 278:189–202.

Pawlik M, Cania B, Thijs S, Vangronsveld J, Piotrowska-Seget Z. 2017. Hydrocarbon degradation potential and plant growth-promoting activity of culturable endophytic bacteria of *Lotus corniculatus* and *Oenothera biennis* from a long-term polluted site. *Environmental Science and Pollution Research* 24(24):19640–19652.

Pawlowska TE, Chaney RL, Chin M, Charvat I.2000. Effects of metal phytoextraction practices on the indigenous community of arbuscular mycorrhizal fungi at a metal-contaminated landfill. *Applied and Environmental Microbiology* 66(6):2526–2530.

Payne AN, DiChristina TJ. 2006. A rapid mutant screening technique for detection of technetium [Tc(VII)] reduction-deficient mutants of *Shewanella oneidensis* MR-1. *FEMS Microbiology Letters* 259(2):282–287.

Pei H, Shao Y, Chanway CP, Hu W, Meng P, Li Z, Chen Y, Ma G. 2016. Bioaugmentation in a pilot-scale constructed wetland to treat domestic wastewater in summer and autumn. *Environ Science and Pollution Research* 23(8):7776–7785.

Peleg Z, Blumwald E. 2011. Hormone balance and abiotic stress tolerance in crop plants. *Current Opinion in Plant Biology* 14(3):290–295.

Peng A, Liu J, Gao Y, Chen Z 2013. Distribution of endophytic bacteria in *Alopecurus aequalis* Sobol and *Oxalis corniculata* L. from soils contaminated by polycyclic aromatic hydrocarbons. *PLOS ONE* 8(12):e83054.

Peuke AD, Rennenberg H. 2005. Phytoremediation. *EMBO Reports* 6(6):497–501.

Phieler R, Voit A, Kothe E. 2014. Microbially supported phytoremediation of heavy metal contaminated soils: Strategies and applications. *Advances in Biochemical Engineering and Biotechnology* 141:211–235.

Philippot L, Raaijmakers J, Lemanceau P et al. 2013. Going back to the roots: The microbial ecology of the rhizosphere. *Nature Reviews in Microbiology* 11:789–799.

Phillips LA, Germida JJ, Farrell RE, Greer CW. 2008. Hydrocarbon degradation potential and activity of endophytic bacteria associated with prairie plants. *Soil Biology and Biochemistry* 40:3054.

Pikovskaya RI. 1948.Mobilization of phosphates in soil in connection with the vital activities of some microbial species. *Mikrobiologiya* 17:362–370.

Płociniczak T, Sinkkonen A, Romantschuk M, Piotrowska-Seget Z. 2013. Characterization of *Enterobacter intermedius* MH8b and its use for the enhancement of heavy metals uptake by *Sinapis alba* L. *Applied Soil Ecology* 63:1–7.

Prapagdee B, Khonsue N. 2015. Bacterial-assisted cadmium phytoremediation by *Ocimum gratissimum* L. in polluted agricultural soil: A field trial experiment. *International Journal of Environmental Science and Technology* 12:3843–3852.

Prashar P, Kapoor N, Sachdeva S. 2014. Rhizosphere: Its structure, bacterial diversity and significance. *Reviews in Environmental Science and Biotechnology* 13:63–77.

Qiao Q, Wang F, Zhang J, Chen Y, Zhang C, Liu G, Zhang H, Ma C, Zhang J. 2017. The variation in the rhizosphere microbiome of cotton with soil type, genotype and developmental Stage. *Scientific Reports* 7:3940.

Qin H, Brookes PC, Xu J. 2014. *Cucurbita* spp. and *Cucumis sativus* enhance the dissipation of polychlorinated biphenyl congeners by stimulating soil microbial community development. *Environmental Pollution* 184:306–312.

Raaijmakers JM, Paulitz TC, Steinberg C, Alabouvette C, Moenne-Loccoz Y. 2009. The rhizosphere: A playground and battlefield for soil borne pathogens and beneficial microorganisms. *Plant and Soil* 321:341–361.

Rabie GH. 2005. Role of arbuscular mycorrhizal fungi in phytoremediation of soil rhizosphere spiked with poly aromatic hydrocarbons. *Mycobiology* 33(1):41–50.

Rajkumar M, Ae N, Freitas H. 2009. Endophytic bacteria and their potential to enhance heavy metal phytoextraction. *Chemosphere* 77(2):153–160.

Rajkumar M, Ae N, Prasad MN, Freitas H. 2010. Potential of siderophore-producing bacteria for improving heavy metal phytoextraction. *Trends in Biotechnology* 28(3):142–149.

Rajkumar M, Freitas H. 2008. Influence of metal resistant-plant growth-promoting bacteria on the growth of *Ricinus communis* in soil contaminated with heavy metals. *Chemosphere* 71(5):834–842.

Rajkumar M, Ma Y, Freitas H. 2008. Characterization of metal-resistant plant-growth promoting *Bacillus weihenstephanensis* isolated from serpentine soil in Portugal. *Joural of Basic Microbiology* 48(6):500–508.

Rajkumar M, Nagendran R, Lee KJ, Lee WH, Kim SZ. 2006. Influence of plant growth promoting bacteria and Cr^{6+} on the growth of Indian mustard. *Chemosphere* 62(5):741–748.

Rajkumar M, Sandhya S, Prasad MNV, Freitas H. 2012. Perspectives of plant-associated microbes in heavy metal phytoremediation. *Biotechnology Advances* 30.

Rajtor M, Piotrowska-Seget Z. 2016. Prospects for arbuscular mycorrhizal fungi (AMF) to assist in phytoremediation of soil hydrocarbon contaminants. *Chemosphere* 162:105–116.

Ramesh A, Sharma SK, Sharma MP, Yadav N, Joshi OP. 2014. Inoculation of zinc solubilizing *Bacillus aryabhattai* strains for improved growth, mobilization and biofortification of zinc in soybean and wheat cultivated in vertisols of central India. *Applied Soil Ecology* 73:87–96.

Ramos JL, Diaz E, Dowling D, deLorenzo V, Molin S, O'Gara F, Ramos C, Timmis KN. 1994. The behavior of bacteria designed for biodegradation. *Biotechnology* 12:1349–1356.

Ranjard L, Nazaret S, Cournoyer B. 2003. Freshwater bacteria can methylate selenium through the thiopurine methyltransferase pathway. *Applied and Environmental Microbiology* 69:3784–3790.

Rascio N, Navari-Izzo F. 2011. Heavy metal hyperaccumulating plants: How and why do they do it? And what makes them so interesting? *Plant Science* 180(2):169–181.

Rashid M, Samina K, Najma A, Sadia A, Farooq L. 2004. Organic acids production and phosphate solubilization by phosphate solubilizing microorganisms under *in vitro* conditions. *Pakistan Journal of Biological Sciences* 7:187–196.

Raskin II, Smith RD, Salt DE. 1997. Phytoremediation of metals: Using plants to remove pollutants from the environment. *Current Opinion in Biotechnology* 8(2):221–226.

Reddy KR, Wright A, Ogram A, DeBusk, WF, Newman S. 2002. Microbial processes regulating carbon cycling in subtropical wetlands. *The 17th World Congress of Soil Science*, Thailand. Symposium no. 11, Paper 982: 1–12.

Redfern LK, Gunsch CK. 2016. Endophytic phytoaugmentation: Treating wastewater and runoff through augmented phytoremediation. *Industrial Biotechnology* 12(2):83–90.

Reed MLE, Glick BR. 2005. Growth of canola (*Brassica napus*) in the presence of plant growth-promoting bacteria and either copper or polycyclic aromatic hydrocarbons. *Canadian Journal of Microbiology* 51:1061–1069.

Rehman K, Imran A, Amin I, Afzal M. 2018. Inoculation with bacteria in floating treatment wetlands positively modulates the phytoremediation of oil field wastewater. *Journal of Hazardous Materials* 349:242–251.

Ren X-M, Guo S-J, Tian W et al. 2019. Effects of plant growth-promoting bacteria (PGPB) inoculation on the growth, antioxidant activity, Cu uptake, and bacterial community structure of rape *(Brassica napus* L.) Grown in Cu-contaminated agricultural soil. *Frontiers in Microbiology* 10:1455.

Rigamonte TA, Pylro VS, Duarte GF. 2010. The role of mycorrhization helper bacteria in the establishment and action of ectomycorrhizae associations. *Brazilian Journal of Microbiology* 41:832–840.

Rillig MC, Wright SF, Evine VT. 2002. The role of arbuscular mycorrhizal fungi and glomalin in soil aggregation: Comparing effects of five plant species. *Plant and Soil* 238(2):325–333.

Roane TM, Pepper IL. 2000. Microorganisms and metal pollution. In: Maier RM, Pepper IL, Gerba CB (eds) *Environmental Microbiology*. London: Academic, p. 55.

Rocha IMV, Silva KNO, Silva DR, Martínez-Huitle CA, Santos EV. 2019. Coupling electrokinetic remediation with phytoremediation for depolluting soil with petroleum and the use of electrochemical technologies for treating the effluent generated. *Separation and Purification Technology* 208(8):194–200.

Rodrigues RR, Moon J, Zhao B, Williams MA. 2017. Microbial communities and diazotrophic activity differ in the root-zone of Alamo and Dacotah switchgrass feedstocks. *GCB Bioenergy* 9:1057–1070.

Rodríguez H, Fraga R. 1999. Phosphate solubilizing bacteria and their role in plant plant growth promotion. *Biotechnology Advances* 17(4–5):319–339.

Rodriguez H, Vessely S, Shah S, Glick BR. 2008. Effect of a nickel-tolerant ACC deaminase-producing pseudomonas strain on growth of nontransformed and transgenic canola plants. *Current Microbiology* 57:170–174.

Rojas-Tapias DF, Bonilla R, Dussán J. 2014. Effect of inoculation and co-inoculation of *Acinetobacter* sp. RG30 and *Pseudomonas putida* GN04 on growth, fitness, and copper accumulation of Maize (*Zea mays*). *Water, Air, & Soil Pollution* 225:2232.

Rojjanateeranaj P, Sangthong C, Prapagdee B. 2017. Enhanced cadmium phytoremediation of *Glycine max* L. through bioaugmentation of cadmium-resistant bacteria assisted by biostimulation. *Chemosphere* 185:764–771.

Román-Ponce B, Reza-vázquez DM, Gutiérrez-Paredes S et al. 2017. Plant growth-promoting traits in rhizobacteria of heavy metal-resistant plants and their effects on *Brassica nigra* seed germination. *Pedosphere* 27:511–526.

Rosenblueth M, Martínez-Romero E. 2006. Bacterial endophytes and their interactions with hosts. *Molecular Plant Microbe Interaction* 19(8):827–137.

Runes HB, Jenkins JJ, Bottomley PJ. 2001. Atrazine degradation by bioaugmented sediment from constructed wetlands. *Applied Microbiology and Biotechnology* 57(3):427–432.

Ryu C-M, Farag MA, Hu C-H, Reddy MS, Wei H-X, Paré PW, Kloepper JW. 2003. Bacterial volatiles promote growth in Arabidopsis. *Proceeding of National Academy of Science USA* 100(8):4927–4932.

Ryu C-M, Hu C-H, Locy RD, Kloepper JW. 2005. Study of mechanisms for plant growth promotion elicited by rhizobacteria in *Arabidopsis thaliana. Plant and Soil* 268(1):285–292.

Saleem H, Arslan M, Rehman K, Tahseen R, Afzal M. 2019. *Phragmites australis*—a helophytic grass—can establish successful partnership with phenol-degrading bacteria in a floating treatment wetland. *Saudi Journal of Biological Sciences* 26(6):1179–1186.

Salt D, Blaylock M, Kumar N et al. 1995. Phytoremediation: A novel strategy for the removal of toxic metals from the environment using plants. *Nature Biotechnology* 13:468–474.

Salt DE, Smith RD, Raskin I. 1998. Phytoremediation. *Plant and Molecular Biology* 49:643–668.

Sandermann H Jr. 1994. Higher plant metabolism of xenobiotics: The 'green liver' concept. *Pharmacogenetics* 4(5):225–241.

Santoyo G, Moreno-Hagelsieb G, Orozco-Mosqueda M, del C, Glick BR. 2016. Plant growth-promoting bacterial endophytes. *Microbiological Research* 183:92–99.

Saravanan VS, Kalaiarasan P, Madhaiyan M, Thangaraju M. 2007. Solubilization of insoluble zinc compounds by *Gluconacetobacter diazotrophicus* and the detrimental action of zinc ion (Zn^{2+}) and zinc chelates on root knot nematode *Meloidogyne incognita. Letters in Applied Microbiology* 44:235–241.

Saxena G, Chandra R, Bharagava RN. 2017. Environmental pollution, toxicity profile and treatment approaches for tannery wastewater and its chemical pollutants. *Reviews in Environmental Contamination and Toxicology* 240:31–69.

Schäffner A, Messner B, Langebartels C, Sandermann H.2002. Genes and enzymes for in-planta phytoremediation of air, water and soil. *Acta Biotechnology* 22(1–2):141–152.

Schell MA. 1985. Transcriptional control of the nah and sal hydrocarbon-degradation operons by the nahR gene product. *Gene* 36:301–309.

Schneider J, Bundschuh J, do Nascimento CWA. 2016. Arbuscular mycorrhizal fungi-assisted phytoremediation of a lead-contaminated site. *Science of the Total Environment* 572:86–97.

Segura A, Rodríguez-Conde S, Ramos C, Ramos JL. 2009. Bacterial responses and interactions with plants during rhizoremediation. *Microbial Biotechnology* 2(4):452–464

Seidel K. 1961. Zur Problematik der Keim- und Pflanzengewasser. *Verh. Internat. Verein. Limnol.* 14:1035–1039.

Selvi A, Rajasekar A, Theerthagiri J, Ananthaselvam A, Sathishkumar K, Madhavan J, Rahman PKSM. 2019. Integrated remediation processes toward heavy metal removal/recovery from various environments-A Review. *Frontiers in Environmental Science* 7:66.

Sessitsch A, Hardoim P, Döring J et al. 2012. Functional characteristics of an endophyte community colonizing rice roots as revealed by metagenomic analysis. *Molecular Plant Microbe Interaction* 25(1):28–36.

Sessitsch A, Kuffner M, Kidd P, Vangronsveld J, Wenzel WW, Fallmann K, Puschenreiter M. 2013. The role of plant-associated bacteria in the mobilization and phytoextraction of trace elements in contaminated soils. *Soil Biology and Biochemistry* 60:182–194.

Shao Y, Pei H, Hu W. 2013. Nitrogen removal by bioaugmentation in constructed wetlands for rural domestic wastewater in autumn. *Desalination and Water Treatment* 51:6624–6631.

Shao Y, Pei H, Hu W, Chanway CP, Meng P, Ji Y, Li Z. 2014. Bioaugmentation in lab scale constructed wetland microcosms for treating polluted river water and domestic wastewater in northern China. *International Biodeterioration & Biodegradation* 95:151–159.

Sharma A, Johri BN, Sharma AK, Glick BR. 2003. Plant growth-promoting bacterium *Pseudomonas* sp. strain GRP3 influences iron acquisition in mung bean (*Vigna radiata* L. Wilzeck). *Soil Biology and Biochemistry* 35(7):887–894.

Sharma S, Tiwari S, Hasan A, Saxena V, Pande LM. 2018. Recent advances in conventional and contemporary methods for remediation of heavy metal-contaminated soils. *3 Biotech* 8:216.

Sharma SB, Sayyed RZ, Trivedi MH, Gobi TA. 2013. Phosphate solubilizing microbes: Sustainable approach for managing phosphorus deficiency in agricultural soils. *Springer Plus* 2:587.

Shehzadi M, Fatima K, Imran A, Mirza MS, Khan, Afzal QMM. 2015. Ecology of bacterial endophytes associated with wetland plants growing in textile effluent for pollutant-degradation and plant growth-promotion potentials. *Plant Biosystems* 150(6):1261–1270.

Shen ZG, Zhao FJ, McGrath SP. 1997. Uptake and transport of zinc in the hyperaccumulator *Thlaspi caerulescens* and the non-hyperaccumulator *Thlaspi ochroleucum*. *Plant, Cell and Environment* 20:898–906.

Sheng X, He L, Wang Q, Ye H, Jiang C. 2008b. Effects of inoculation of biosurfactant-producing *Bacillus* sp. J119 on plant growth and cadmium uptake in a cadmium-amended soil. *Journal of Hazardous Materials* 155:17–22.

Sheng X, Xia J, Jiang C, He L, Qian M. 2008a. Characterization of heavy metal-resistant endophytic bacteria from rape (*Brassica napus*) roots and their potential in promoting the growth and lead accumulation of rape. *Environmental Pollution* 156:1164–1170.

Sheng XF, Chen XB, He LY. 2008c. Characteristics of an endophytic pyrene-degrading bacterium of *Enterobacter* sp. 12J1 from *Allium macrostemon* Bunge. *International Biodeterioration & Biodegradation* 62(2):88–95.

Sheng XF, Jiang CY, He LY. 2008d. Characterization of plant growth-promoting *Bacillus edaphicus* NBT and its effect on lead uptake by Indian mustard in a lead-amended soil. *Canadian Journal of Microbiology* 54(5):417–422.

Sheng XF, Xia JJ. 2006. Improvement of rape (*Brassica napus*) plant growth and cadmium uptake by cadmium-resistant bacteria. *Chemosphere* 64(6):1036-42.

Shidore T, Dinse T, Ohrlein J, Becker A, Reinhold-Hurek B. 2012. Transcriptomic analysis of responses to exudates reveal genes required for rhizosphere competence of the endophyte *Azoarcus* sp. strain BH72. *Environmental Microbiology* 14:2775–2787.

Shim H, Chauhan S, Ryoo D, Bowers K, Thomas SM, Canada KA, Burken JG, Wood TK. 2000. Rhizosphere competitiveness of trichloroethylene-degrading, poplar-colonizing recombinant bacteria. *Applied and Environmental Microbiology* 66:4673–4678.

Shin W, Islam R, Benson A et al. 2016. Role of diazotrophic bacteria in biological nitrogen fixation and plant growth improvement. *Korean Journal of Soil Science and Fertilizer* 49(1):17–29.

Siciliano SD, Fortin N, Mihoc A et al. 2001. Selection of specific endophytic bacterial genotypes by plants in response to soil contamination. *Applied and Environmental Microbiology* 67:2469–2475.

Siciliano SD, Germida JJ. 1998a. Degradation of chlorinated benzoic acid mixtures by plant–bacteria associations. *Environmental Toxicology and Chemistry* 17(4):728–733.

Siciliano SD, Germida JJ 1998b. Mechanisms of phytoremediation: Biochemical and ecological interactions between plants and bacteria. *Environmental Reviews* 6:65–79.

Singh AK, Cameotra SS. 2013. Efficiency of lipopeptide biosurfactants in removal of petroleum hydrocarbons and heavy metals from contaminated soil. *Environmental Science and Pollution Research International* 20(10):7367–7376.

Singh BK, Walker A. 2006. Microbial degradation of organophosphorus compounds. *FEMS Microbiological Reviews* 30:428–471.

Singh G, Pankaj U, Chand S, Verma RK. 2019. Arbuscular mycorrhizal fungi-assisted phytoextraction of toxic metals by *Zea mays* L. from tannery sludge. *Soil and Sediment Contamination: An International Journal* 28(8).

Sinha S, Mukherjee SK. 2008. Cadmium-induced siderophore production by a high Cd-resistant bacterial strain relieved Cd toxicity in plants through root colonization. *Current Microbiology* 56(1):55–60.

Siripan O, Thamchaipenet A, Surat W. 2018. Enhancement of the efficiency of Cd phytoextraction using bacterial endophytes isolated from *Chromolaena odorata*, a Cd hyperaccumulator. *International Journal of Phytoremediation* 20(11):1096–1105.

Soleimani M, Hajabbasi MA, Afyuni M, Mirlohi A, Borggaard OK, Holm PE. 2010. Effect of endophytic fungi on cadmium tolerance and bioaccumulation by *Festuca arundinacea* and *Festuca pratensis*. *International Journal of Phytoremediation* 12(6):535–549.

Spruyt A, Buck MT, Mia A, Straker CJ. 2014. Arbuscular mycorrhiza (AM) status of rehabilitation plants of mine wastes in South Africa and determination of AM fungal diversity by analysis of the small subunit rRNA gene sequences. *South African Journal of Botany* 94:231–237.

Sriprang R, Hayashi M, Yamashita M, Ono H, Saeki K, Murooka Y. 2002. A novel bioremediation system for heavy metals using the symbiosis between leguminous plant and genetically engineered rhizobia. *Journal of Biotechnology* 99(3):279–293.

Srivastava S, Singh N. 2014. Mitigation approach of arsenic toxicity in chickpea grown in arsenic amended soil with arsenic tolerant plant growth promoting *Acinetobacter* sp. *Ecological Engineering* 70:146–153.

Srivastava S, Verma PC, Chaudhry V, Singh N, Abhilash PC, Kumar KV, Sharma N, Singh N. 2013. Influence of inoculation of arsenic-resistant *Staphylococcus arlettae* on growth and arsenic uptake in *Brassica juncea* (L.) Czern. Var. R-46. *Journal of Hazardous Materials* 262:1039–1047.

Steenhoudt O, Vanderleyden J. 2000. *Azospirillum*, a free-living nitrogen-fixing bacterium closely associated with grasses: Genetic, biochemical and ecological aspects. *FEMS Microbiol Reviews* 24(4):487–506.

Stottmeister U, Wiessner A, Kuschk P et al. 2003. Effects of plants and microorganisms in constructed wetlands for wastewater treatment. *Biotechnology Advances* 22(1–2):93–117.

Suman J, Uhlik O, Viktorova J, Macek T. 2018. Phytoextraction of Heavy Metals: A promising tool for clean-up of polluted environment? *Frontiers in Plant Science* 9:1476.

Sun K, Liu J, Jin L, Gao Y. 2014. Utilizing pyrene-degrading endophytic bacteria to reduce the risk of plant pyrene contamination. *Plant and Soil* 374(1–2):251–262.

Sun LN, Zhang YF, He LY, Chen ZJ, Wang QY, Qian M, Sheng XF. 2010. Genetic diversity and characterization of heavy metal-resistant-endophytic bacteria from two copper-tolerant plant species on copper mine wasteland. *Bioresour Technology* 101(2):501–509.

Sun M, Fu D, Teng Y, Shen Y, Luo Y, Li Z, Christie P. 2011. In situ phytoremediation of PAH-contaminated soil by intercropping alfalfa (*Medicago sativa* L.) with tall fescue (*Festuca arundinacea* Schreb.) and associated soil microbial activity. *Journal of Soils and Sediments* 11:980–989.

Taghavi S, Barac T, Greenberg B, Borremans B, Vangronsveld J, van der Lelie D. 2005. Horizontal gene transfer to endogenous endophytic bacteria from poplar improves phytoremediation of toluene. *Applied and Environmental Microbiology* 71(12):8500–8505.

Taghavi S, Garafola C, Monchy S et al. 2009. Genome survey and characterization of endophytic bacteria exhibiting a beneficial effect on growth and development of poplar trees. *Applied And Environmental Microbiology* 75(3):748–757.

Taghavi S, Weyens N, Vangronsveld J, van der Lelie D. 2011. Improved phytoremediation of organic contaminants through engineering of bacterial endophytes of trees. In: Pirttilä AM, Frank AC (eds) *Endophytes of Forest Trees*. Netherlands: Springer, pp. 205–216.

Taiolia E, Sram RJ, Garte S, Kalina I, Popove TA, Farmer PB. 2007. Effects of polycyclic aromatic hydrocarbons (PAHs) in environmental pollution on exogenous and oxidative DNA damage (EXPAH project): Description of the population under study. *Mutation Research/Fundamental and Molecular Mechanisms of Mutagenesis* 620(1–2):1–6.

Taniguchi M, Kiba T, Sakakibara H, Ueguchi C, Mizuno T, Sugiyama T. 1998. Expression of Arabidopsis response regulator homologs is induced by cytokinins and nitrate. *FEBS Letters* 429:259–262.

Tara N, Iqbal M, Khan QM, Afzal M. 2019. Bioaugmentation of floating treatment wetlands for the remediation of textile effluent. *Water and Environment Journal* 33:124–134

Tarkka MT, Frey-Klett P. 2008. Mycorrhiza helper bacteria. In: Varma A (ed.) *Mycorrhiza*. Berlin, Heidelberg: Springer.

Tarkka MT, Herrmann S, Wubet T et al. 2013. OakContigDF159.1, a reference library for studying differential gene expression in Quercus robur during controlled biotic interactions: Use for quantitative transcriptomic profiling of oak roots in ectomycorrhizal symbiosis. *New Phytologist* 199:529–540.

Tchounwou PB, Yedjou CG, Patlolla AK, Sutton DJ. 2012. Heavy metals toxicity and the environment. *EXS* 101:133–164.

Temmink RJM, Harpenslager SF, Smolders AJP, van Dijk G, Peters RCJH, Lamers LPM, van Kempen MML. 2018. *Azolla* along a phosphorus gradient: Biphasic growth response linked to diazotroph traits and phosphorus-induced iron chlorosis. *Scientific Reports* 8:4451.

Teng Y, Luo Y, Sun X, Tu C, Xu L, Liu W, Li Z, Christie P. 2010. Influence of arbuscular mycorrhiza and rhizobium on hytoremediation by alfalfa of an agricultural soil contaminated with weathered PCBs: A Field Study. *International Journal of Phytoremediation* 12(5):516–533.

Thijs S, Sillen W, Weyens N, Vangronsveld J. 2017. Phytoremediation: State-of-the-art and a key role for the plant microbiome in future trends and research prospects. *International Journal of Phytoremediation* 19(1):23–38.

Thijs S, Van Dillewijn P, Sillen W et al. 2014. Exploring the rhizospheric and endophytic bacterial communities of *Acer pseudoplatanus* growing on a TNT-contaminated soil: Towards the development of a rhizocompetent TNT-detoxifying plant growth promoting consortium. *Plant and Soil* 385:15–36.

Tirry N, Tahri Joutey N, Sayel H, Kouchou A, Bahafid W, Asri M, El Ghachtouli N. 2018. Screening of plant growth promoting traits in heavy metals resistant bacteria: Prospects in phytoremediation. *Journal of Genetic Engineering and Biotechnology* 16(2):613–619.

Tiwari S, Singh SN, Garg SK. 2012. Stimulated phytoextraction of metals from fly ash by microbial interventions. *Environmental Technology* 33(19–21):2405–2413.

Tomasino SF, Leister RT, Dimock MB, Beach RM, Kelly JL. 1995. Field performance of *Clavibacter xyli* subsp. cynodontis expressing the insecticidal protein gene cryIA. of *Bacillus thuringiensis* against European corn borer in field corn. *Biological Control* 5:442–448.

Turnau K, Jurkiewicz A, Lingua G, Barea JM, Gianinazzi-Pearson V. 2005. Role of arbuscular mycorrhiza and associated microorganisms in phytoremediation of heavy metal-polluted sites. In: MNV P, Sajwan KS, Naidu R (eds) *Trace Elements in the Environment. Biogeochemistry, Biotechnology, and Bioremediation*. Boca Raton, London, New York: CRC Taylor & Francis, pp. 235–225.

Ullah A, Heng S, Munis MFH, Fahad S, Yang X. 2015. Phytoremediation of heavy Phytoremediation of heavy metals assisted by plant growth promoting (PGP) bacteria: A review. *Environmental and Experimental Botany* 17:28–40.

Ullah A, Mushtaq H, Ali H, Munis MF, Javed MT, Chaudhary HJ. 2014. Diazotrophs-assisted phytoremediation of heavy metals: A novel approach. *Environmental Science and Pollution Research International* 22(4):2505–2514.

United States Environmental Protection Agency (USEPA). 1999. Use of monitored natural attenuation at superfund, RCRA corrective action, and underground storage tank sites, directive number 9200.4-17P.

United States Environmental Protection Agency (USEPA). 2000. Ground water issue. EPA.

United States Environmental Protection Agency (USEPA). 2006. In situ and ex situ biodegradation technologies for remediation of contaminated sites. EPA/625/R-06/015.

United States Environmental Protection Agency (USEPA). 2012 A citizen guide to bioremediation. EPA 542-F-12-003.

Vacheron J, Desbrosses G, Bouffaud ML et al. 2013. Plant growth-promoting rhizobacteria and root system functioning. *Frontiers in Plant Science* 4:356.

van Aken B, Peres CM, Doty SL, Yoon JM, Schnoor JL. 2004a. Methylobacterium populi sp. nov., a novel aerobic, pink-pigmented, facultatively methylotrophic, methane-utilizing bacterium isolated from poplar trees (Populus deltoides x nigra DN34). *International Journal of Systematic and Evolutionary Microbiology* 54:1191–1196.

van Aken B, Tehrani R, Schnoor JL. 2011. Endophyte-assisted phytoremediation of explosives in poplar trees by Methylobacterium populi BJ001 T. In: Pirttila AM, Frank AC (eds) *Endophytes of Forest Trees: Biology and Applications*. London: Springer, pp. 217–236.

van Aken B, Yoon JM, Schnoor JL et al. 2004b. Biodegradation of nitro-substituted explosives 2,4,6-Trinitrotoluene, Hexahydro-1,3,5-Trinitro-1,3,5- Triazine, and Octahydro-1,3,5,7-Tetranitro-1,3,5,7-Tetrazocine by a Phytosymbiotic Methylobacterium sp. associated with poplar tissues (*Populus deltoidesxnigra* DN34). *Applied and Environmental Microbiology* 70:508–517.

van der Heijden MGA, Bardgett RD, Van Straalen NM. 2008. The unseen majority: Soil microbes as drivers of plant diversity and productivity in terrestrial ecosystems. *Ecology Letters* 11:296–310.

van der Lelie D, Taghavi S, Monchy S et al. 2009. Poplar and its Bacterial Endophytes: Coexistence and Harmony. *Critical Reviews in Plant Sciences* 28:346–358.

Vangronsveld J. 2009. Bioaugmentation with engineered endophytic bacteria improves contaminant fate in phytoremediation. *Environmental Science and Technology* 43:9413–9418.

van Loon LC. 2007. Plant responses to plant growth-promoting rhizobacteria. *European Journal of Plant Pathology* 119:243–254.

van Loon LC, Bakker PAHM. 2005. Induced systemic resistance as a mechanism of disease suppression by rhizobacteria. In: Siddiqui ZA (ed.) *PGPR: Biocontrol and Biofertilization*. Dordrecht, The Netherlands: Springer.

Vannini C, Carpentieri A, Salvioli A et al. . 2016. An interdomain network: The endobacterium of a mycorrhizal fungus promotes antioxidative responses in both fungal and plant hosts. *New Phytologist* 211(1):265–275.

Van Peer R, Niemann GJ, Schippers B. 1991. Induced resistance and phytoalexin accumulation in biological control of fusarium wilt of carnation by *Pseudomonas* sp. strain WCS417r. *Phytopathology* 91:728–34.

Vansuyt G, Robin A, Briat JF, Curie C, Lemanceau P. 2007. Iron acquisition from Fe-pyoverdine by *Arabidopsis thaliana*. *Molecular Plant Microbe Interaction* 20:441–447.

Vassilev N, Vassileva M, Nikolaeva I. 2006. Simultaneous P-solubilizing and biocontrol activity of microorganisms: Potentials and future trends. *Applied Microbiology and Biotechnology* 71:137–144.

Venkatesh NM, Vedaraman N. 2012. Remediation of soil contaminated with copper using Rhamnolipids produced from *Pseudomonas aeruginosa* MTCC 2297 using waste frying rice bran oil. *Annals of Microbiology* 62:85–91.

Vespermann A, Kai M, Piechulla B. 2007. Rhizobacterial volatiles affect the growth of fungi and Arabidopsis thaliana. *Applied and Environmental Microbiology* 73(17):5639–5641.

Villacieros M, Whelan C, Mackova M, Molgaard J, Sánchez-Contreras M, Lloret J, Aguirre de Cárcer D, Oruezábal RI, Bolanos L, Macek T, Karlson U, Dowling DN, Martín M, Rivilla R. 2005. Polychlorinated biphenyl rhizoremediation by *Pseudomonas fluorescens* F113 derivatives, using a *Sinorhizobium meliloti* nod system to drive bph gene expression. *Applied and Environmental Microbiology* 71:2687–2694.

Vleesschauwer D, Höfte M. 2009. Rhizobacteria-induced systemic resistance. *Advances in Botanical Research* 51:223–281.

Vymazal J. 2001. Types of constructed wetlands for wastewater treatment: Their potential for nutrient removal. In: Vymazal J (ed.) *Transformations of Nutrients in Natural and Constructed Wetlands*. Leiden, The Netherlands: Backhuys Publishers, pp. 1–93.

Vymazal J. 2007. Removal of nutrients in various types of constructed wetlands. *Science of the Total Environment* 380:48–65.

Vymazal J. 2010. Constructed wetlands for wastewater treatment. *Water* 2:530–549.

Vymazal J. 2011. Constructed wetlands for wastewater treatment: Five decades of experience. *Environmental Science & Technology* 45(1):61–69.

Vymazal J, Greenway M, Tonderski K, Brix H, Mander Ü. 2006. Constructed wetlands for wastewater treatment. In: Verhoeven JTA, Beltman B, Bobbink R, Whigham DF (eds) *Wetlands and Natural Resource Management*. Ecological Studies (Analysis and Synthesis), vol 190. Berlin, Heidelberg: Springer.

Wagner SC. 2011. Biological nitrogen fixation. *Nature Education Knowledge* 3(10):15.

Wakelin SA, Warren RA, Harvey PR, Ryder MH. 2004. Phosphate solubilization by Penicillium spp. closely associated with wheat roots. *Biology and Fertility of Soils* 40:36–43.

Wang F, Lin X, Yin R. 2005. Heavy metal uptake by arbuscular mycorrhizas of *Elsholtzia splendens* and the potential for phytoremediation of contaminated soil. *Plant and Soil* 269(1/2):225–232.

Wang FY, Lin XG, Yin R. 2007. Effect of arbuscular mycorrhizal fungal inoculation on heavy metal accumulation of maize grown in a naturally contaminated soil. *International Journal of Phytoremediation* 9(4):345–353.

Wang P, Marsh EL, Ainsworth EA, Leakey ADB, Sheflin AM, Schachtman DP. 2017. Shifts in microbial communities in soil, rhizosphere and roots of two major crop systems under elevated CO_2 and O_3. *Scientific Reports* 7:15019.

Wang Q, Liu J, Zhu H. 2018. Genetic and molecular mechanisms underlying symbiotic specificity in legume-rhizobium interactions. *Frontiers in Plant Science* 9:313.

Wang Q, Xiong D, Zhao P, Yu X, Tu B, Wang G. 2011. Effect of applying an arsenic resistant and plant growth–promoting rhizobacterium to enhance soil arsenic phytoremediation by *Populus deltoides* LH05-17. *Journal of Applied Microbiology* 111:1065–1074.

Wang S, Guo S, Li F et al. 2016. Effect of alternating bioremediation and electrokinetics on the remediation of n-hexadecane-contaminated soil. *Scientific Reports* 6:23833.

Wang X, Zhang X, Liu X, Huang Z et al. 2019. Physiological, biochemical and proteomic insight into integrated strategies of an endophytic bacterium *Burkholderia cenocepacia* strain YG-3 response to cadmium stress. *Metallomics* 11(7):1252–1264.

Wang Y, H Li, W Zhao, X He, J Chen, X Geng, M Xiao 2010. Induction of toluene degradation and growth promotion in corn and wheat by horizontal gene transfer within endophytic bacteria. *Soil Biology and Biochemistry* 42(7):1051–1057.

Wani PA, Khan MS. 2013. Nickel detoxification and plant growth promotion by multi metal resistant plant growth promoting *Rhizobium* species RL9. *Bulletin of Environmental Contamination and Toxicology* 91(1):117–124.

Wei G, Kloepper JW, Tuzun S. 1991. Induction of systemic resistance of cucumber to *Colletotrichum orbiculare* by select strains of plant growth-promoting rhizobacteria. *Phytopathology* 81:1508–1512.

Wei Y, Chen Z, Wu F, Li J, ShangGuan Y, Li F, Zeng QR, Hou H. 2015. Diversity of arbuscular mycorrhizal fungi associated with a Sb accumulator plant, ramie (*Boehmeria nivea*), in an active Sb Mining. *Journal of Microbiology and Biotechnology* 25(8): 1205–1215.

Wei Z, Hu X, Li X et al. 2017 The rhizospheric microbial community structure and diversity of deciduous and evergreen forests in Taihu Lake area, China. *PLOS ONE* 12(4):e0174411.

Weyens N, Croes S, Dupae J, Newman L, van der Lelie D, Carleer R, Vangronsveld J. 2010a. Endophytic bacteria improve phytoremediation of Ni and TCE co-contamination. *Environmental Pollution* 158:2422–2427.

Weyens N, Schellingen K, Beckers B et al. 2013. Potential of willow and its genetically engineered associated bacteria to remediate mixed Cd and toluene contamination. *Journal of Soil and Sediments* 13:176–188.

Weyens N, Taghavi S, Barac T et al. 2009a. Bacteria associated with oak and ash on a TCE-contaminated site: Characterization of isolates with potential to avoid evapotranspiration of TCE. *Environmental Science and Pollution Research International* 16:830–843.

Weyens N, Truyens S, Dupae J et al. 2010b. Potential of the TCE-degrading endophyte *Pseudomonas putida* W619-TCE to improve plant growth and reduce TCE phytotoxicity and evapotranspiration in poplar cuttings. *Environmental Pollution (Barking, Essex : 1987)* 158:2915–2919.

Weyens N, Truyens S, Saenen E et al. 2011. Endophytes and their potential to deal with co-contamination of organic contaminants (toluene) and toxic metals (nickel) during phytoremediation. *International Journal of Phytoremediation* 13(3):244–255.

Weyens N, van der Lelie D, Taghavi S, Newman L, Vangronsveld J. 2009b. Exploiting plant-microbe partnerships to improve biomass production and remediation. *Trends in Biotechnology* 27:591–598.

Weyens N, van der Lelie D, Taghavi S, Vangronsveld J. 2009c. Phytoremediation: Plant endophyte partnerships take the challenge. *Current Opinion in Biotechnology* 20:248–254.

Weyens W, Vangronsveld N, Newman JL. 2009d. Poplar and its bacterial endophytes: Coexistence and harmony. *Critical Reviews in Plant Science* 28:346–358.

Whiting SN, de Souza MP, Terry N 2001. Rhizosphere bacteria mobilize Zn for hyperaccumulation by *Thlaspi caerulescens*. *Environmental Science and Technology* 35(15):3144–3150.

Wick LY, Shi L, Harms H. 2007. Electro-bioremediation of hydrophobic organic soil-contaminants: A review of fundamental interactions. *Electrochimica Acta* 52(10):3441–3448.

Woese CR, Fox GE. 1977. Phylogenetic structure of the prokaryotic domain: The primary kingdoms. *Proceeding of National Academy of Science USA* 74(11):5088–5090.

Wu CH, Bernard SM, Andersen GL, Chen W. 2009. Developing microbe–plant interactions for applications in plant-growth promotion and disease control, production of useful compounds, remediation and carbon sequestration. *Microbial Biotechnology* 2(4):428–440.

Wu CH, Wood TK, Mulchandani A, Chen W. 2006a. Engineering plant-microbe symbiosis for rhizoremediation of heavy metals. *Applied and Environmental Microbiology* 72(2):1129–1134.

Wu SC, Cheung KC, Luo YM, Wong MH. 2006b. Effects of inoculation of plant growth-promoting rhizobacteria on metal uptake by *Brassica juncea*. *Environmental Pollution* 140(1):124–135.

Wuana RA, Okieimen FE. 2011. Heavy metals in contaminated soils: A review of sources, chemistry, risks and best available strategies for remediation. *ISRN Ecology* 201:20. Article ID 402647.

Xiao Y, Wang X, Chen W, Huang Q. 2017. Isolation and identification of three potassium-solubilizing bacteria from rape rhizospheric soil and their effects on ryegrass. *Geomicrobiology Journal* 34(10):873–880.

Xin G, Glawe D, Doty SL. 2009. Characterization of three endophytic indole-3-acetic acid-producing yeasts occurring in Populus trees. *Mycological Research* 113:973–980.

Xun F, Xie B, Liu S, Guo C. 2015. Effect of plant growth-promoting bacteria (PGPR) and arbuscular mycorrhizal fungi (AMF) inoculation on oats in saline-alkali soil contaminated by petroleum to enhance phytoremediation. *Environmental Science and Pollution Research* 22:598–608.

Yadav SK, Juwarkar AA, Balki AB et al. 2014. Microorganism-assisted phytoremediation of heavy metal and endosulfan contaminated soil. *Reviews on Environmental Health* 29(1–2):41–42.

Yan F, Reible D. 2015. Electro-bioremediation of contaminated sediment by electrode enhanced capping. *Journal of Environmental Management* 155:154–161.

Yang X, Feng Y, He Z, Stoffella PJ. 2005. Molecular mechanisms of heavy metal hyperaccumulation and phytoremediation. *Journal of Trace Elements in Medicine and Biology* 18(4):339–353.

Yang Y, He C, Huang L, Ban Y, Tang M. 2017. The effects of arbuscular mycorrhizal fungi on glomalin-related soil protein distribution, aggregate stability and their relationships with soil properties at different soil depths in lead-zinc contaminated area. *PLOS ONE* 12(8):e0182264.

Yang Y, Liang Y, Ghosh A, Song Y, Chen H, Tang M. 2015. Assessment of arbuscular mycorrhizal fungi status and heavy metal accumulation characteristics of tree species in a lead-zinc mine area: Potential applications for phytoremediation. *Environmental Science and Pollution Research International* 22(17):13179–13193.

Yang Y, Liang Y, Han X, Chiu TY, Ghosh A, Chen H, Tang M. 2016. The roles of arbuscular mycorrhizal fungi (AMF) in phytoremediation and tree-herb interactions in Pb contaminated soil. *Scientific Reports* 6:20469.

Yaseen DA, Scholz M. 2019. Textile dye wastewater characteristics and constituents of synthetic effluents: A critical review. *International Journal of Environmental Science and Technology* 16(2):1193–1226.

Ye S, Zeng G, Wu H, Zhang C, Dai J, Liang J, Yu J, Ren X, Yi H, Cheng M, Zhang C. 2017. Biological technologies for the remediation of co-contaminated soil. *Critical Reviews in Biotechnology* 37(8):1062–1076.

Yee DC, Maynard JA, Wood TK. 1998. Rhizoremediation of trichloroethylene by a recombinant, root-colonising *Pseudomonas fluorescens* stain expressing toluene ortho-monooxygenase constituitively. *Applied and Environmental Microbiology* 64:112–118.

Ying G-G, Rai S. 2003. Kookana degradation of five selected endocrine-disrupting chemicals in seawater and marine sediment. *Environmental Science & Technology* 37(7):1256–1260.

Yong X, Chen Y, Liu W, Xu L, Zhou J, Wang S, Chen P, Ouyang P, Zheng T. 2014. Enhanced cadmium resistance and accumulation in *Pseudomonas putida* KT2440 expressing the phytochelatin synthase gene of *Schizosaccharomyces pombe*. *Letters in Applied Microbiology* 58(3):255–261.

Yoon J, Cao X, Zhou Q, Ma LQ. 2006. Accumulation of Pb, Cu, and Zn in native plants growing on a contaminated Florida site. *Science of the Total Environment* 368(2):456–464.

Yousaf S, Afzal M, Reichenauer TG, Brady CL, Sessitsch A. 2011. Hydrocarbon degradation, plant colonization and gene expression of alkane degradation genes by endophytic Enterobacter ludwigii strains. *Environmental Pollution* 159:2675–2683.

Yousaf S, Andria V, Reichenauer TG, Smalla K, Sessitsch A. 2010. Phylogenetic and functional diversity of alkane degrading bacteria associated with Italian ryegrass (*Lolium multiflorum*) and Birdsfoot trefoil (*Lotus corniculatus*) in a petroleum oil-contaminated environment. *Journal of Hazardous Materials* 184(1–3):523–532.

Yuan M, He H, Xiao L, Zhong T, Liu H, Li S, Deng P, Ye Z, Jing Y. 2014. Enhancement of Cd phytoextraction by two *Amaranthus* species with endophytic *Rahnella* sp. JN27. *Chemosphere* 103:99–104.

Zaidi A, Khan MS, Ahemad M, Oves M. 2009. Plant growth promotion by phosphate solubilizing bacteria. *Acta Microbiologica et Immunologica Hungarica* 56(3):263–284.

Zazouli MA, Mahdavi Y, Bazrafshan E, Balarak D. 2014. Phytodegradation potential of bisphenolA from aqueous solution by *Azolla filiculoides*. *Journal of Environmental Health Science and Engineering* 12:66.

Zhang C-B, Liu W-L, Pan X-C, Guan M, Liu S-Y, Ge Y, Chang J. 2014. Comparison of effects of plant and biofilm bacterial community parameters on removal performances of pollutants in floating island systems. *Ecological Engineering* 73:58–63.

Zhang H, Kim M-S, Krishnamachari V, Payton P, Sun Y, Grimson M, Farag MA, Ryu C-M, Allen R, Melo IS. 2007. Rhizobacterial volatile emissions regulate auxin homeostasis and cell expansion in *Arabidopsis*. *Planta* 226 (4):839.

Zhang H, Xie X, Kim MS, Kornyeyev DA, Holaday S, Paré PW. 2008. Soil bacteria augment Arabidopsis photosynthesis by decreasing glucose sensing and abscisic acid levels in planta. *Plant Journal* 56(2):264–273.

Zhang M, Guo S, Li F, Wu B. 2017. Distribution of ion contents and microorganisms during the electro-bioremediation of petroleum-contaminated saline soil. *Journal of Environmental Science Health, Part A Toxic Hazardous Substances and Environmental Engineering* 52(12):1141–1149.

Zhang Y, Chen F-S, Wu X-Q et al. 2018. Isolation and characterization of two phosphate-solubilizing fungi from rhizosphere soil of moso bamboo and their functional capacities when exposed to different phosphorus sources and pH environments. *PLOS ONE* 13(7):e0199625.

Zhang YF, He LY, Chen ZJ, Wang QY, Qian M, Sheng XF. 2011. Characterization of ACC deaminase-producing endophytic bacteria isolated from copper-tolerant plants and their potential in promoting the growth and copper accumulation of *Brassica napus*. *Chemosphere* 83(1):57–62.

Zheng BX, Ibrahim M, Zhang DP, Bi QF, Li HZ, Zhou GW, Ding K, Peñuelas J, Zhu YG, Yang XR. 2018. Identification and characterization of inorganic-phosphate-solubilizing bacteria from agricultural fields with a rapid isolation method. *AMB Express* 8:47.

Zhou JL. 1999. Zn biosorption by Rhizopus arrhizus and other fungi. *Applied Microbiology and Biotechnology* 51:686–693.

Zhu X, Ni X, Liu J, Gao Y. 2013. Application of endophytic bacteria to reduce persistent organic pollutants contamination in plants. *Clean Soil Air Water* 42(3):306–310.

Index